Blast Mitigation for Structures

1999 Status Report on the DTRA/TSWG Program

Committee for Oversight and Assessment of Blast Effects and Related Research
Board on Infrastructure and the Constructed Environment
Commission on Engineering and Technical Systems
National Research Council

NATIONAL ACADEMY PRESS
WASHINGTON, D.C.

NATIONAL ACADEMY PRESS 2101 Constitution Avenue, N.W. Washington, D.C. 20418

NOTICE: The project that is the subject of this report was approved by the Governing Board of the National Research Council, whose members are drawn from the councils of the National Academy of Sciences, the National Academy of Engineering, and the Institute of Medicine. The members of the committee responsible for the report were chosen for their special competencies and with regard for appropriate balance.

This report has been reviewed by a group other than the authors according to procedures approved by a Report Review Committee consisting of members of the National Academy of Sciences, the National Academy of Engineering, and the Institute of Medicine.

This study was supported by Contract No. DSWA01-98-C-0075 between the National Academy of Sciences and the Defense Threat Reduction Agency. Any opinions, findings, conclusions, or recommendations expressed in this publication are those of the authors and do not necessarily reflect the view of the organizations or agencies that provided support for this project.

Copyright 2000 by the National Academy of Sciences. All rights reserved.

Available from:
Board on Infrastructure and the Constructed Environment
National Research Council
2101 Constitution Avenue, NW
Washington, DC, 20418
Printed in the United States of America.

National Academy of Sciences
National Academy of Engineering
Institute of Medicine
National Research Council

The **National Academy of Sciences** is a private, nonprofit, self-perpetuating society of distinguished scholars engaged in scientific and engineering research, dedicated to the furtherance of science and technology and to their use for the general welfare. Upon the authority of the charter granted to it by the Congress in 1863, the Academy has a mandate that requires it to advise the federal government on scientific and technical matters. Dr. Bruce M. Alberts is president of the National Academy of Sciences.

The **National Academy of Engineering** was established in 1964, under the charter of the National Academy of Sciences, as a parallel organization of outstanding engineers. It is autonomous in its administration and in the selection of its members, sharing with the National Academy of Sciences the responsibility for advising the federal government. The National Academy of Engineering also sponsors engineering programs aimed at meeting national needs, encourages education and research, and recognizes the superior achievements of engineers. Dr. William A. Wulf is president of the National Academy of Engineering.

The **Institute of Medicine** was established in 1970 by the National Academy of Sciences to secure the services of eminent members of appropriate professions in the examination of policy matters pertaining to the health of the public. The Institute acts under the responsibility given to the National Academy of Sciences by its congressional charter to be an adviser to the federal government and, upon its own initiative, to identify issues of medical care, research, and education. Dr. Kenneth I. Shine is president of the Institute of Medicine.

The **National Research Council** was organized by the National Academy of Sciences in 1916 to associate the broad community of science and technology with the Academy's purposes of furthering knowledge and advising the federal government. Functioning in accordance with general policies determined by the Academy, the Council has become the principal operating agency of both the National Academy of Sciences and the National Academy of Engineering in providing services to the government, the public, and the scientific and engineering communities. The Council is ad-ministered jointly by both Academies and the Institute of Medicine. Dr. Bruce M. Alberts and Dr. William A. Wulf are chairman and vice chairman, respectively, of the National Research Council.

www.national-academies.org

COMMITTEE FOR OVERSIGHT AND ASSESSMENT OF BLAST EFFECTS AND RELATED RESEARCH

METE A. SOZEN, *chair*, Purdue University, West Lafayette, Indiana
STEPHEN W. ATTAWAY, Sandia National Laboratories, Albuquerque, New Mexico
ERIK AUF DER HEIDE, Agency for Toxic Substances and Disease Registry, Atlanta, Georgia
W. GENE CORLEY, Construction Technology Laboratories, Skokie, Illinois
EVE HINMAN, Hinman Consulting Engineers, Inc., San Francisco, California
ROBERT P. KENNEDY, RPK Structural Mechanics Consulting, Escondido, California
SAM A. KIGER, University of Missouri-Columbia, Columbia
STUART L. KNOOP, Oudens and Knoop, Architects, PC, Chevy Chase, Maryland
JOHANNA LAPIERRE, RTKL Associates, Inc., Washington, D.C.
MARK LOIZEAUX, Controlled Demolition Inc., Phoenix, Maryland
J. L. MERRITT, J.L. Merritt Consulting Engineer, Yucaipa, California
DAVID J. PELGRIM, E. K. Fox & Associates, Ltd., Fairfax, Virginia
EUGENE SEVIN, consultant, Lyndhurst, Ohio
CHARLES H. THORNTON, Thornton/Tomasetti Engineers, New York, New York

Staff

RICHARD G. LITTLE, director, Board on Infrastructure and the Constructed Environment
LORI JACKSON, administrative associate

BOARD ON INFRASTRUCTURE AND THE CONSTRUCTED ENVIRONMENT

RICHARD N. WRIGHT, *chair,* National Institutes of Standards and Technology (retired), Gaithersburg, Maryland
GREGORY BAECHER, University of Maryland, College Park
JONATHAN BARNETT, urban planner, Washington, D.C.
MAX BOND, Davis, Brody, Bond, LLP, New York, New York
JACK E. BUFFINGTON, University of Arkansas, Fayetteville
MARY COMERIO, University of California, Berkeley
CLAIRE FELBINGER, American University, Washington, D.C.
PAUL GILBERT, Parsons, Brinckerhoff, Quade, and Douglas, Seattle, Washington
AMY GLASMEIER, Pennsylvania State University, University Park
CHRISTOPHER M. GORDON, Massachusetts Port Authority, East Boston
NEIL GRIGG, Colorado State University, Fort Collins
JEREMY ISENBERG, Weidlinger Associates, New York, New York
MARTHA A. ROZELLE, The Rozelle Group, Ltd., Phoenix, Arizona
DAVID SKIVEN, General Motors Corporation, Detroit, Michigan
SARAH SLAUGHTER, Massachusetts Institute of Technology, Cambridge
ERIC TEICHOLZ, Graphic Systems, Inc., Cambridge, Massachusetts
RAE ZIMMERMAN, New York University, New York, New York

Staff

RICHARD G. LITTLE, director, Board on Infrastructure and the Constructed Environment
LYNDA L. STANLEY, executive director, Federal Facilities Council
JOHN A. WALEWSKI, project officer
LORI JACKSON, administrative associate
GAIL KELLEY, research assistant

Acknowledgments

This report has been reviewed in draft form by individuals chosen for their diverse perspectives and knowledge of the subject matter, in accordance with procedures approved by the NRC Report Review Committee. The purpose of this independent review is to provide candid and critical comments that will assist the NRC in making this report as sound as possible and to ensure that it meets institutional standards for objectivity, evidence, and responsiveness to the study charge. The review comments and draft manuscript remain confidential to protect the integrity of the deliberative process. We wish to thank the following individuals for their participation in the review of this report:

Chester A. Canada, U.S. Department of Defense
David R. Dibner, Bernard Johnson, Inc. (retired)
William J. Hall, University of Illinois
Theodor Krauthammer, Pennsylvania State University
Joseph Penzien, International Civil Engineering Consultants
Robin M. Wagner, Agency for Toxic Substances and Disease Registry
Richard M. Wright, National Institutes of Standards and Technology (retired)

While these individuals provided constructive comments and suggestions, it must be emphasized that responsibility for the final content of the report rests with the authoring committee and the NRC.

ACKNOWLEDGMENTS

Contents

	EXECUTIVE SUMMARY	1
1	INTRODUCTION	6
	Scope of the Study	7
	Organization of the Study	7
	Organization of the Report	8
	Reference	8
2	THE BLAST MITIGATION FOR STRUCTURES PROGRAM	9
	Program Review	9
	Program Assessment	11
	Conclusion and Recommendation	14
	References	14
3	REVIEW OF PROGRAM ACTIVITIES	15
	Structural Issues	15
	Reducing Injuries through Nonstructural Approaches	23
	International Programs	26
	Conclusions and Recommendations	26
	References	28
4	TECHNOLOGY TRANSFER	31
	Background	31
	Needs of the Engineering Community	31
	Building Codes and Standards	32
	Role of Academia	34
	Handling of Sensitive Information	34
	A Framework for Technology Transfer	35
	Conclusion and Recommendation	36
	References	36
	APPENDIXES	
A	Biographies of Committee Members	41
B	Blast Mitigation for Structures, Program Master Plan	45

List of Figures and Tables

FIGURES

2-1	Relative funding levels for Task 2 subtasks and work-unit categories	12
2-2	Framework for defining and refining the BMSP	12
4-1	Model of a technology transfer strategy	36

TABLES

2-1	Funding Levels by Major Component	10
2-2	Funding by Fiscal Year ($000's)	10
2-3	Task 2 Subtasks and Work-Unit Categories	11

Acronyms

ACI	American Concrete Institute
ASCE	American Society of Civil Engineers
ATC	Applied Technology Council
BFRL	Building and Fire Research Laboratory
BMAG	Blast Mitigation Action Group
BMSP	Blast Mitigation for Structures Program
BSSC	Building Seismic Safety Council
COTS	commercial-off-the-shelf
CTS-1	Controlled Test Structure-1
DoD	U.S. Department of Defense
DTRA	Defense Threat Reduction Agency
EERI	Earthquake Engineering Research Institute
ERDC	Engineering Research and Development Center (U.S. Army Corps of Engineers)
FEMA	Federal Emergency Management Agency
FY	fiscal year
LBTF	large blast and thermal simulator
NEHRP	National Earthquake Hazard Reduction Program
NIST	National Institute of Standards and Technology
NRC	National Research Council
NSF	National Science Foundation
SEAOC	Structural Engineers Association of California
TSWG	Technical Support Working Group
V&V	verification and validation (of computational codes)

Executive Summary

The Blast Mitigation for Structures Program (BMSP) is a research and development activity conducted by the Defense Threat Reduction Agency (DTRA) to improve the performance of buildings that are targets of terrorist attack. The primary goal of the BMSP is to reduce loss of life and injuries to the occupants of these buildings through the development of innovative techniques for new structures and retrofitting existing facilities. The program encompasses the analysis, computational modeling, and physical testing of buildings and components, and structural and nonstructural systems. A report in 1995 by the National Research Council (NRC), *Protecting Buildings From Bomb Damage: Transfer of Blast-Effects Mitigation Technology from Military to Civilian Applications,* recommended that a research and development program be focused on the mitigation of blast effects and the transfer of relevant technology (NRC, 1995). Prior to the publication of that report, the NRC had published two studies, *The Embassy of the Future,* a series of recommendations for the design of future U.S. embassies (NRC, 1986), and *The Protection of Federal Office Buildings Against Terrorism,* a guide for federal agencies to improve the security of persons, buildings, and information from terrorist attack (NRC, 1988). Because of the NRC's expertise and long-standing involvement in this area, DTRA requested that the NRC review and assess the BMSP's implications for engineering, architectural, and building practices, disaster preparedness, rescue and recovery operations, and emergency medical services, and recommend appropriate research topics and a strategy for technology transfer.

In response to that request, the NRC established the Committee for Oversight and Assessment of Blast-Effects and Related Research under the auspices of the Board on Infrastructure and the Constructed Environment. The committee was asked to:

- Assist in the development of a blast-effects research agenda and provide recommendations for activity priorities. This will include assessing the scope and focus of related, on-going research, both in this country and internationally, to assure that efforts are well-integrated; evaluating the capability of the existing research infrastructure to achieve research objectives; and determining the possible need for a national test facility to carry out the research program.
- Recommend appropriate mechanisms to achieve effective transfer of research results and existing technologies to civilian government agencies and commercial engineering and architectural practice;
- Develop recommendations for outreach and knowledge dissemination activities to be undertaken by DTRA and other agencies;
- Review and comment on proposed curriculum or training materials designed to enable civilian engineers and architects to apply the principles of protective design and analysis to civilian buildings and other constructed facilities.
- Provide a forum to enhance interaction and information sharing among other stakeholder government agencies such as the General Services Administration, Federal Emergency Management Agency, U.S. Army Corps of Engineers, Centers for Disease Control and Prevention, Bureau of Alcohol, Tobacco, and Firearms, Department of Transportation, Department of State, etc., and state and local governments.

The committee's objective is to monitor and assist an evolving program that has activities subject to ongoing revision. DTRA anticipates that the BMSP will be funded through fiscal year 2003 (FY03) at an annual level of approximately $10 million and that the program could potentially be continued beyond 2003. Because of the ongoing and long-term nature of the BMSP, in this first annual review the committee makes some broad programmatic observations and recommendations for improving DTRA's strategic planning, priority setting, and resource allocation over the life of the program.

The committee's principal findings and recommendations are based on an analysis of current program plans, descriptions of work units, and budget allotments. Although the BMSP includes most of the activities the committee identified as necessary for a comprehensive program to improve the likelihood of survival of occupants of buildings subject to terrorist bombing, the committee did identify a number of specific areas where increased emphasis could bring immediate benefits and some current activities that should be reduced or redirected. In subsequent reports, the committee will evaluate in more detail how well the program objectives are being met and will suggest reallocations of the resources that will become available if the committee's recommendations have been implemented. The chapters in this report include additional observations and guidance on specific issues.

CONCLUSIONS AND RECOMMENDATIONS

Conclusion 1. The overall plan of the BMSP is appropriately focused on the explicit, and laudable, goal of protecting human life. Nevertheless, the BMSP would be improved by initiating a formalized strategic planning process for identifying and reaching consensus on knowledge gaps, reassessing them in light of lessons learned through individual program activities or studies outside the BMSP, refining or establishing new objectives, and identifying the activities that should be continued, initiated, or abandoned.

Recommendation 1. The Defense Threat Reduction Agency should allocate sufficient time and resources to formalize a strategic planning process for reviewing and refining the Blast Mitigation for Structures Program on an annual basis. Planning of the next full-scale building test should be delayed until a strategic plan has been developed that defines the functions of the analytical and experimental components of the program in terms of overall program goals, and with respect to one another. The plan should also identify and establish pathways for developing the questions that may only be answered by large-scale building tests.

Conclusion 2. Although the Program Master Plan includes many activities that could yield worthwhile benefits, the committee identified several modifications to the BMSP to be considered in the next program cycle.

Recommendation 2a. All analytical and experimental activities should be designed to test a specific hypothesis about the outcome. With respect to full-scale tests, parametric studies should be conducted to determine what could be learned from the test on the basis of the proposed instrumentation.

Recommendation 2b. The program should take full advantage of the advances in parallel-processor computing made by the U.S. Department of Defense and the U.S. Department of Energy to improve the capability and ease of use of computational tools for predicting structural responses to bomb blasts.

Recommendation 2c. The residual strength of blast-damaged structural components should be investigated more fully. For example, tests of full scale columns representative of buildings ten stories and more should be included, as well as a series of tests to evaluate how well common rebar splices and connections can function after being damaged by blasts.

Recommendation 2d. The Blast Mitigation for Structures Program should consider conducting a series of tests on masonry structures, including tests of unreinforced masonry for benchmarking purposes and tests of a range of reinforcement techniques to improve protection. A series of tests on construction typical of long-span buildings should also be considered.

Recommendation 2e. The Blast Mitigation for Structures Program should place a higher priority on the development and evaluation of retrofitting techniques—particularly on creative conceptual retrofitting measures that would prevent a life-threatening progressive collapse following a blast.

Recommendation 2f. The Blast Mitigation for Structures Program should focus more attention on the behavior of nonstructural systems in the blast environment, including (1) tests of the effectiveness of various types of interior partitions or perimeter zones of "soft" space in protecting occupants and contents, and (2) comparisons of floor-based systems of mechanical and electrical distribution and typical overhead systems.

Recommendation 2g. The Blast Mitigation for Structures Program should evaluate the key factors affecting the ease and rapidity with which trapped or injured occupants can be extricated from damaged buildings and whether rescuers can safely enter areas of the collapsed structure to render aid. In cooperation with urban search and rescue teams, the program should support simulated rescue and recovery operations to refine or improve rescue techniques.

Conclusion 3. The design and engineering approaches favored by the industrial contractors and government laboratories that are implementing the BMSP may be more appropriate to traditional military and defense objectives despite the emphasis of the BMSP on nonstructural solutions, injury reduction, and improved rescue and recovery techniques.

Recommendation 3. The contractor base should be broadened to increase the representation of the commercial architectural and engineering communities, as well as specialists in injury prevention, disaster medicine, and technology transfer, particularly in the planning phases of the program.

Conclusion 4. The committee is in complete agreement with the BMSP's emphasis on determining progressive-collapse vulnerability of buildings in selected attack scenarios but believes this ability would be improved by fuller coordination of research activities.

Recommendation 4. The Defense Threat Reduction Agency should adopt a balanced approach toward understanding and preventing the progressive collapse of buildings. This approach should include coordinated physical testing, experimentation, and analyses and should guide the planning of research activities and the interpretation and synthesis of the results.

Conclusion 5. Full-scale testing of structural systems has been overemphasized at this relatively early stage of the program at the expense of reduced-scale testing, the development of retrofitting techniques for existing buildings, the testing of nonstructural building systems, and the investigation of technologies related to injury prevention.

Recommendation 5. The Defense Threat Reduction Agency should not construct another full-scale test structure until the results of previous experiments on Controlled Test Structure-1 (CTS-1) have been fully analyzed and understood. At this stage of the program, DTRA should rely more on experiments with scaled elements and scaled assemblies of elements wherever scale effects are well understood.

Conclusion 6. Controlled Test Structure-1 has been underutilized so far; although it has been damaged in previous tests, it still has considerable value for testing full-scale structural components and nonstructural elements.

Recommendation 6. Controlled Test Structure-1 should not be tested to failure because it can still be used as a reaction frame for component tests.

Conclusion 7. Although, the inventory of existing buildings vulnerable to blast damage far exceeds the number of new buildings that will be constructed in the foreseeable future, the BMSP appears to have placed more emphasis on methods applicable to new construction than on retrofitting techniques for existing structures.

Recommendation 7. The development of tools for conducting vulnerability assessments and strengthening existing buildings should be given a higher priority. Resources should also be allocated to investigating construction techniques that permit the rapid rehabilitation of blast-damaged buildings.

Conclusion 8. The Blast Mitigation for Structures Program has a unique opportunity to determine how requirements and techniques for earthquake-resistant designs could apply to blast-resistant designs, as well as to identify and assess design features and materials that could improve building performance over a range of hazards (e.g.., earthquake, fire, flood, and extreme wind) that could impact the safety of the occupants.

Recommendation 8. The Blast Mitigation for Structures Program should incorporate activities with the maximum potential for multihazard mitigation. Because design features that provide multihazard resistance are likely to generate more interest among designers and manufacturers than design features that promise only blast resistance, multihazard features could ultimately reduce the cost and increase the application of improved building practices and products.

Conclusion 9. Data on blast-related injuries and building damage are limited and, therefore, have hindered the development of statistically valid damage-prediction and epidemiological models.

Recommendation 9. The Blast Mitigation for Structures Program should initiate an institutionalized process that can be quickly mobilized for collecting critical data related to blast damage and injuries in buildings that are subject to bomb damage.

Conclusion 10. The barriers to the complete and effective transfer of the results of the BMSP will require considerable time and effort to overcome. A convenient way to reduce the transfer time would be to use existing institutional infrastructures (i.e., building code and standards-writing organizations, professional and technical organizations, universities, and research centers) to disseminate knowledge.

Recommendation 10. A workshop to develop a road map for transferring technology for mitigating blast effects should be scheduled as soon as possible. To assist in the ongoing dissemination of information, the Blast Mitigation for Structures Program should consider sponsoring an annual or biennial conference devoted to all aspects of blast-mitigation design, engineering, injury prevention, and rescue and recovery.

REFERENCES

NRC (National Research Council). 1986. The Embassy of the Future: Recommendations for the Design of Future U.S. Embassy Buildings. Washington, D.C.: National Academy Press.

NRC. 1988. The Protection of Federal Office Buildings Against Terrorism. Washington, D.C.: National Academy Press.

NRC. 1995. Protecting Buildings from Bomb Damage: Transfer of Blast-Effects Mitigation Technologies from Military to Civilian Applications. Washington, D.C.: National Academy Press.

1

Introduction

In the past six years, the United States and its citizens have been the victims of five major bombing attacks. The attack on the World Trade Center in New York City in February 1993 and on the Alfred P. Murrah Building in Oklahoma City in April 1995 occurred on U.S. soil. Attacks off shore included bombing of the military complex at Dhahran, Saudi Arabia, in 1996 and the bombings of the U.S. embassies in Kenya and Tanzania in 1998. These events, previous bombings in Kuwait and Beirut, and lethal attacks elsewhere have generated considerable concern about the ability of the United States to protect buildings and their occupants from bombings and other direct physical attacks.

In 1995, in the wake of the bombings of the World Trade Center and the Murrah Federal Building, the National Research Council (NRC) published *Protecting Buildings From Bomb Damage: Transfer of Blast-Effects Mitigation Technologies from Military to Civilian Applications* (NRC, 1995). This report found that much of the structural research and testing that had been done in support of military missions during the Cold War was generally applicable to civilian design practice and recommended that a program be initiated to continue research and technology transfer to improve the performance of civilian buildings, minimize casualties, and improve rescue and recovery operations in cases of terrorist bombing attacks.

Although designing and building structures to withstand the effects of explosive devices has been a topic of active interest and research in the defense community for many years, these activities were focused mostly on preventing structural failures that would compromise or destroy mission-critical systems. This focus led to the construction of heavily reinforced bunker-type structures and underground facilities that are fundamentally different from the commercial civilian facilities used extensively to house military troops and civilian personnel. Since the bombings in Oklahoma City and Saudi Arabia (Khobar Towers), which both caused many casualties, force protection (i.e., ensuring the safety of personnel in all types of structures) has become a critical mission parameter for the U.S. military.

In 1997, at the direction of the Congress, the U.S. Department of Defense (DoD) established the Blast Mitigation for Structures Program (BMSP) to identify design and construction practices and improved materials for more robust construction to protect its forces from terrorist attacks. However, this knowledge of blast effects on structures and their subsystems, appropriately refined and applied to commercial design and construction practices, could also save civilian lives and reduce property damage. For example, studies of recent attacks on buildings both here and abroad indicate that the number of fatalities is very strongly correlated with building collapse. However, even if a building does not collapse, blast-induced debris, smoke, the destruction of fire- and life-safety systems, and obstructions to prompt rescue are significant causes of injury and death. If the robustness of the structure and critical systems can be improved, the recovery and reuse of the building would be a major payback of the costs of using blast-mitigating designs. Therefore, it would be of benefit to military and civilian communities to develop an integrated research program to minimize casualties in the event of bomb blasts, as well as facilitating rescue and recovery operations. The research program should also include the planning and provision of emergency medical services in the aftermath of the event.

The Defense Threat Reduction Agency (DTRA), under the sponsorship of the DoD Technical Support Working Group (TSWG), has initiated the BMSP to meet these objectives. To ensure that the research program addresses high-priority needs in a technically sound and cost-effective manner, the former Defense Special Weapons Agency (since subsumed into DTRA) requested that the NRC review the BMSP program annually and offer ongoing recommendations both for conducting the research and for transferring technology to the military and civilian sectors.

SCOPE OF THE STUDY

In response to that request, the NRC assembled an independent panel of experts, the Committee for Oversight and Assessment of Blast-Effects and Related Research, under the auspices of the Board on Infrastructure and the Constructed Environment. The committee was asked to perform the following tasks:

(1) Assist in the development of a blast-effects research agenda and provide recommendations for activity priorities. This will include assessing the scope and focus of related, on-going research, both in this country and internationally, to assure that efforts are well-integrated; evaluating the capability of the existing research infrastructure to achieve research objectives; and determining the possible need for a national test facility to carry out the research program.
(2) Recommend appropriate mechanisms to achieve effective transfer of research results and existing technologies to civilian government agencies and commercial engineering and architectural practice;
(3) Develop recommendations for outreach and knowledge dissemination activities to be undertaken by DTRA and other agencies;
(4) Review and comment on proposed curriculum or training materials designed to enable civilian engineers and architects to apply the principles of protective design and analysis to civilian buildings and other constructed facilities.
(5) Provide a forum to enhance interaction and information sharing among other stakeholder government agencies such as the General Services Administration, Federal Emergency Management Agency, U.S. Army Corps of Engineers, Centers for Disease Control and Prevention, Bureau of Alcohol, Tobacco, and Firearms, Department of Transportation, Department of State, etc., and state and local governments.

ORGANIZATION OF THE STUDY

The 14 members of the committee have expertise in blast-effects research and testing, structural analysis and design, architectural and interior design, seismic safety, disaster preparedness and consequence management, emergency medical services, computer-based modeling and assessment techniques, building code development, and knowledge transfer. Biographical information about the committee members is provided in Appendix A.

The committee held three meetings—in Albuquerque, New Mexico, in May 1999 and in Washington, D.C., in June and September 1999. The committee chair and two committee members also attended a program review organized by DTRA in August 1999. The committee drew heavily on briefings provided by the sponsor and its contractors, as well as the considerable experience of committee members, to develop the conclusions and recommendations included in this report.

ORGANIZATION OF THE REPORT

The succeeding chapters in this report are organized to address the committee's charge. Chapter 2 contains a review and assessment of the objectives and scope of the BMSP, the committee's assessment of progress to date, and recommendations for improving the efficiency and value of the program. Chapter 3 is a discussion of program elements and the committee's recommendations for specific activities for the next program cycle. Chapter 4 contains the committee's initial observations and recommendations on a strategy for technology transfer.

REFERENCE

NRC (National Research Council). 1995. Protecting Buildings from Bomb Damage: Transfer of Blast-Effects Mitigation Technologies from Military to Civilian Applications. Washington, D.C.: National Academy Press.

2

The Blast Mitigation for Structures Program

PROGRAM REVIEW

Program Goal and Objectives

The overall goal of the BMSP is "to protect people inside buildings from terrorist bomb attacks" (DTRA, 1999). The program seeks to achieve this goal by focusing on two primary areas: (1) reducing the likelihood of structural collapse and (2) minimizing the hazard from flying debris.[1] Specific program objectives are listed below (DTRA, 1999):

- Establish tri-service and interagency collaboration on the program.
- Develop cost-effective methods to retrofit existing structures to mitigate the effects of blast.
- Develop design guidance for new construction to mitigate the effects of blast.
- Test and evaluate commercial-off-the-shelf (COTS) products for their capability to increase the resistance of a structure to blast effects.
- Develop industry standard models for the prediction of blast effects on structures and generate computer modules for use by government and industry.
- Define and produce joint service/agency design and assessment tools (to be defined by the user community).
- Develop simplified injury prediction models so that the benefits of blast mitigation design and retrofits can be evaluated in terms of human injury avoidance.

Program Activities

To implement the BMSP, DTRA has issued numerous contracts and work orders for analyses, computational modeling, and reduced-scale and full-scale physical experiments. Physical testing has been carried out primarily at two sites, White Sands, New Mexico, and the U.S. Army Engineering Research and Development Center (ERDC) in Vicksburg, Mississippi. At White Sands, DTRA Field Command has constructed a full-scale prototype of a concrete flat-slab structure called Controlled Test Structure-1 (CTS-1) to test windows, walls, and structural elements under realistic threat conditions (i.e., the blast effects of large vehicle bombs [equivalent to ~5,000 pounds of TNT]). The large blast and thermal simulator (LBTS), a very large shock tube originally constructed at White Sands to simulate the effects of nuclear

---

[1] The committee believes that the direct effects of explosions on the human body are a potentially significant cause of injuries and death. However, the committee also believes the emphasis of the BMSP on secondary effects (i.e., the response of the building and its subsystems to blast) and their potential to cause injury and death is well focused given the state of knowledge regarding secondary blast effects.

weapons, has proven to be valuable for simulating conventional blast effects on complete curtain-wall assemblies for public and commercial buildings.

A complementary program of reduced-scale and component testing is being conducted at the ERDC Structures Laboratory, which has a long history of model and component testing under blast conditions. The ERDC has also organized the Blast Mitigation Action Group (BMAG) to identify and evaluate commercially available (commercial-off-the-shelf or COTS) products. The BMAG web site provides sources of blast mitigation products and services for new construction and retrofitting existing structures (BMAG, 2000). The BMSP has also made extensive use of private contractors with expertise in structural engineering, structural dynamics, and computational modeling to design experiments and analyze their results.

Program Organization and Funding Profiles

The three primary "tasks" of the BMSP are: (1) technology and construction evaluation; (2) technology development and application; and (3) computer support, technology transfer, administration. Approximately 11 per cent of program funds are unallocated as "reserves." Total program funding for fiscal years 1999 (FY99) through FY03 is anticipated to be $54 million. Funding in FY99 was $8.3 million; funding for FY00 through FY02 is projected to remain level at $12 million, and funding for FY03 is expected to drop to $10 million. DTRA, the parent organization of the BMSP, anticipates that the program will continue beyond FY03 at an annual funding level of $8 million to $10 million. The overall task structure and funding profiles of the BMSP are summarized in Tables 2-1 and 2-2.

TABLE 2-1 Funding Levels by Major Component

Task		Percentage of Funding
1	technology and construction evaluation	8 %
2	technology development and application	64 %
3	computer support, technology transfer, administration[a]	17 %
	Reserves[b]	11%

[a]Includes operations and maintenance support for contributing military services
[b]Unallocated funds
Source: DTRA, 1999b.

TABLE 2-2 Funding by Fiscal Year ($000's)

Task	FY99	FY00	FY01	FY02	FY03	FY04
1	$894	$1,246	$825	$750	$625	$4,340
2	$5,529	$8,914	$8,915	$6,724	$4,650	$34,732
3	$1,880	$1,840	$1,840	$1,845	$1,845	$9,250
Reserves			$420	$2,681	$2,880	$5,981
Totals	$8,303	$12,000	$12,000	$12,000	$10,000	$54,303

Source: DTRA, 1999b.

Task 2, the principal technology-development component of the program, consists of four subtasks, which involve a number of work-unit categories (groups of work units) (see Table 2-3). Figure 2-1 shows relative funding levels for the Task 2 subtasks for FY99 through FY03.

TABLE 2-3 Task 2 Subtasks and Work-Unit Categories

Subtask 2A: Loading Definition (11 % of Task 2 funding)

Work Unit	Percentage of Subtask Funding		
Blast wave propagation	36%	Blast wall effectiveness	64%

Subtask 2B: Structural Collapse (62 % of Task 2 funding)

Work Unit	Percentage of Subtask Funding
Flat-slab construction (CTS-1)[a]	21 %
Load-bearing wall construction	9%
Roof slabs	7%
Blast-resistant construction (CTS-2)[a]	12%
Seismic construction (CTS-3)[a]	12%
Steel-frame construction (CTS-4)[a]	18%
Other structures	21%

[a]Denotes a full-scale field test

Subtask 2C: Debris Hazards (25 % of Task 2 funding)

Work Unit	Percentage of Subtask Funding
Windows and doors	16%
Exterior walls	51%
Building internals	14%
Injury studies	19%

Subtask 2D: Internal Detonations (2 % of Task 2 funding)
Source: DTRA, 1999b.

PROGRAM ASSESSMENT

The committee was asked to determine whether the BMSP Master Plan, as prepared by DTRA, provided sufficient guidance for the research and development elements of the program (DTRA, 1999). To aid in this determination, the committee superimposed the BMSP on a basic logic diagram for program development (i.e., statement of purpose, project identification, priority setting, and implementation) as shown in Figure 2-2. The committee also assessed the extent and balance of the technical activities in the Program Plan. The six major activities, defined by the committee in ascending order of cost and time (but not necessarily value), are listed below:

- review of literature (lowest cost)
- analysis and calculation
- reduced-scale testing
- full-scale component testing
- computational modeling
- full-scale structural testing (highest cost)

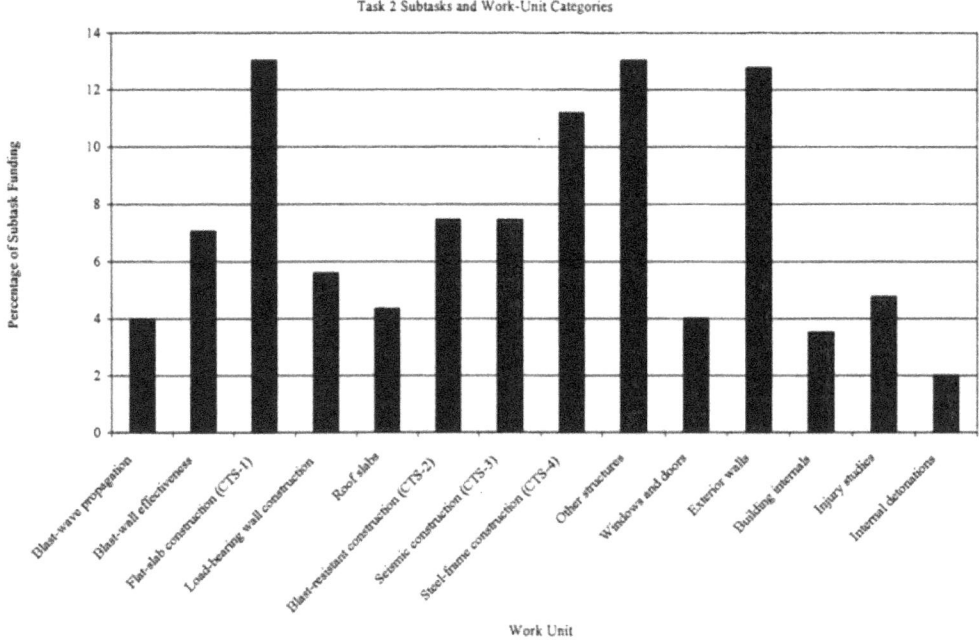

FIGURE 2-1 Relative funding levels for Task 2 subtasks and work-unit categories.

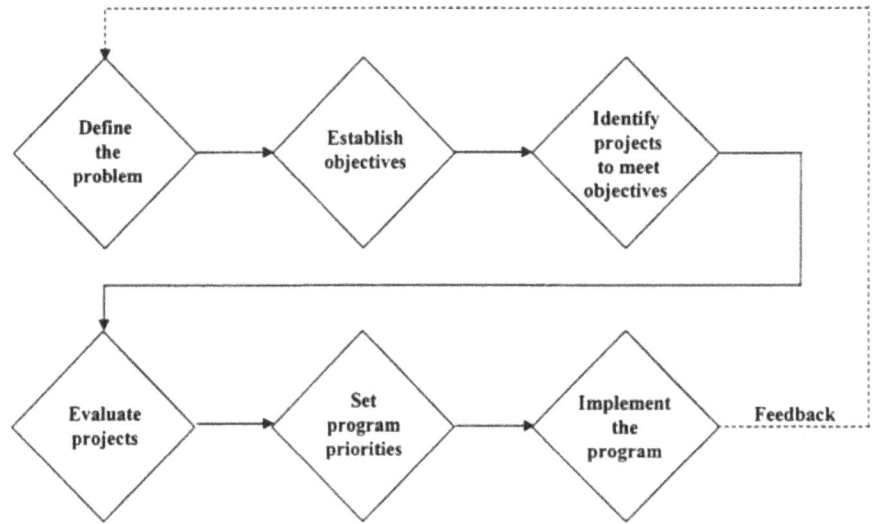

FIGURE 2-2 Framework for defining and refining the BMSP.

Observations

At the higher program levels (i.e., program purpose and focus areas), the committee believes that the BMSP program is focused correctly on protecting people first and physical assets second. Epidemiological studies after the bombing of the Alfred P. Murrah Building in Oklahoma City have shown that building collapse was the primary cause of death and that flying debris was the primary cause of nonfatal injuries in that event (Mallonee, et al., 1996). Studies of the attacks on the Khobar Towers complex and the U.S. embassy in Nairobi also suggest that, in the absence of structural collapse, flying debris was the primary cause of death and injury (U.S. Department of State, 1999). Based on these findings, the BMSP has focused its efforts on reducing both the probability of progressive collapse and the production of hazardous debris.

The committee notes, however, that the program has not developed a formalized strategic planning process for identifying knowledge gaps and assessing them in light of what has been learned by individual program activities, as well as outside the BMSP. A formalized planning process would enable DTRA to refine objectives or establish new ones and determine, on an annual basis, which activities should be continued, initiated, or abandoned. Annual reviews wouldensure that the program remains sufficiently robust to resolve problems that will take several years of study and sufficiently flexible to address emerging issues. The committee believes annual assessments are a key element in the development of a successful program and recommends that DTRA allocate the time and resources to put in place a continuing process for reviewing and refining the BMSP.

The committee also believes that questions must be resolved at an appropriate level of inquiry. In other words, if a short review of the literature and one or two physical experiments at reduced scale will satisfy a particular hypothesis at a 90–95 percent confidence level, there is no need to carry out full-scale experiments. However, some structural phenomena (e.g., phenomena involving the effects of gravity) can only be satisfactorily modeled at full scale or under very sophisticated experimental conditions. The committee believes that the BMSP Master Plan does not yet provide sufficient rationale and guidance to suggest the appropriate analytical or experimental approach on a consistent basis. For this reason the committee recommends that all proposed activities be guided by clearly stated objectives and working hypotheses.

Ultimately, the behavior of a full-scale, fully loaded structure in a high-intensity blast environment may be the only acceptable demonstration of proof of concept—in a political, if not an engineering context. However, full-scale testing is very high in value (and cost) in the experimental hierarchy. Therefore, the practical equivalent of live-fire testing should only be undertaken on full-scale structures as the culmination and validation of a chain of supporting analyses and experiments. Full-scale structural tests should not be used to gain new knowledge but to confirm what has been learned and demonstrate measures that should work in similar conditions.

Although the BMSP includes most of the activities the committee identified as necessary for a comprehensive program to improve the likelihood of survival of occupants of buildings subject to terrorist bombing, the committee did identify a number of specific areas where increased emphasis could bring immediate benefits and some current activities that should be reduced or redirected. In subsequent reports, the committee will evaluate in more detail how well the program objectives are being met and will suggest reallocations of the resources that will become available if the committee's recommendations have been implemented. In the following chapter, the committee addresses technical activities and project-specific aspects of the BMSP and recommends modifications and activities that DTRA should consider for the next program cycle.

CONCLUSION AND RECOMMENDATION

Conclusion 1. The overall plan of the BMSP is appropriately focused on the explicit, and laudable, goal of protecting human life. Nevertheless, the BMSP would be improved by initiating a formalized strategic planning process for identifying and reaching consensus on knowledge gaps, reassessing them in light of lessons learned through individual program activities or studies outside the BMSP, refining or establishing new objectives, and identifying the activities that should be continued, initiated, or abandoned.

Recommendation 1. The Defense Threat Reduction Agency should allocate sufficient time and resources to formalize a strategic planning process for reviewing and refining the Blast Mitigation for Structures Program on an annual basis. Planning of the next full-scale building test should be delayed until a strategic plan has been developed that defines the functions of the analytical and experimental components of the program in terms of overall program goals, and with respect to one another. The plan should also identify and establish pathways for developing the questions that may only be answered by large-scale building tests.

REFERENCES

BMAG (Blast Mitigation Action Group). 2000. BMAG. Available on line at: *http://bmag.nwo.usace.army.mil*

DTRA (Defense Threat Reduction Agency). 1999a. Blast Mitigation for Structures Program Master Plan, June, 1999. Alexandria, Va.: Defense Threat Reduction Agency.

DTRA. 1999b. Blast Mitigation for Structures Program Review, August, 1999. Alexandria, Va.: Defense Threat Reduction Agency.

Mallonee, S., S. Shariat, G. Stennies, R. Waxweiler, D. Hogan, and F. Jordan. 1996. Physical injuries and fatalities resulting from the Oklahoma City bombing. Journal of the American Medical Association 5: 382–387.

U.S. Department of State, 1999. Report of the Accountability Review Boards on the Embassy Bombings in Nairobi and Dar es Salaam on August 7, 1998, January 1999. Available on line at: *http://www.state.gov/www/regions/africa/accountability_report.html*

3

Review of Program Activities

Before commenting on specific technical areas, the committee wishes to point out one important programmatic issue DTRA should address as the program matures. At present, industrial contractors and government laboratories, drawn almost exclusively from DTRA's protective-construction and blast-hardening design community, are implementing the BMSP. Therefore, although this group has first-rate technical capabilities, the committee recommends that the contractor base be expanded to include the commercial architectural and engineering communities, as well as specialists in injury prevention, disaster medicine, and technology transfer, especially for the planning phases of the program. The committee's concern is that, as a group, the existing corps of contractors may be concerned mostly with traditional military and defense objectives, which may not reflect the emphasis of the BMSP on nonstructural solutions, injury reduction methods, and improved rescue and recovery techniques. In addition, although some of the BMSP's contractors are active in key organizations involved in the development of building codes and commercial engineering design, many of them are not. As a result, the BMSP may not be benefiting from complementary developments in the commercial sector. The committee considers this a significant, but easily remedied, problem and strongly recommends that DTRA broaden its contractor base.

STRUCTURAL ISSUES

Progressive Collapse

Progressive structural collapse is a principal, if not the leading, cause of injury and death in building failures, regardless of the source of the loading (e.g., bomb, earthquake, internal explosion). For this reason, predicting and designing to prevent the progressive collapse of a building under a specified attack scenario is (and should be) a primary objective of the BMSP. After considering whether the study of progressive collapse should be addressed through physical testing or computation and analysis, the committee concluded that both are necessary.

This dual experimental approach could investigate the behavior of complete structures or structural components. However, because buildings are complex systems that can have large variances between design specifications and as-built conditions, a test structure may not accurately mimic the progressive failure of a real building. Although an experienced engineer can often estimate the likelihood that a specific building will collapse by superimposing a damage scenario on the design, because of the variances described above, the actual progress of a collapse is essentially a stochastic process. For example, following major earthquakes, nonstructural building components, such as mechanical piping, partition walls, equipment, heavy-duty storage facilities (shelving and file cabinets), and curtain walls that can transfer some of the dead load to lower levels, have been observed to keep buildings standing that would otherwise have been expected to fail (Loizeaux, 1999). This phenomenom, (i.e., a complete progressive failure that is just barely contained) has been observed in Mexico, California, Japan, Guam, and more recently in Turkey and Taiwan. Unfortunately, although nonstructural elements obviously have an important role to play in determining individual outcomes, their random contribution to preventing a collapse cannot be easily included in structural models. (Random factors are discussed in more detail later in this chapter.)

Because of fiscal realities, only a limited number of full-scale tests can be conducted. For this reason, the committee believes that testing complete, full-scale structures is not a practical way to gather knowledge about progressive collapse. Nevertheless, experimental data to improve the collapse-resistance of buildings is important. Therefore, reinforcement details in columns, beams, and floors and their interconnections can be tested at the component level for various blast loadings, and the information gained can then be used to determine additional measures for minimizing the extent of collapse. The committee believes that physical testing of components at both full scale and reduced scale is a cost-effective means of adding valuable information to the knowledge base for reducing the likelihood of progressive collapse and is a valid area of investigation for the BMSP. The committee recommends that basic research into progressive collapse also be supported at academic institutions— either through the BMSP directly or through BMSP's support of other cooperative arrangements.

Computational Modeling

Advances in parallel computing have increased both the size and speed of computational tools (on the order of 10^2 to 10^3), and consequently, computational modeling is changing rapidly. As the speed of central processing units continues to increase, hardware is outpacing software, which in turn is outpacing the availability of experimental data for validating models. Despite this computational capability, attempts to develop an accurate model of progressive collapse have been unsuccessful, however. In the explosives demolition business (which provides many opportunities to compare predictions with observed results under relatively controlled conditions), predemolition predictions for a structural frame using computational models do not generally compare well with the results obtained under actual field conditions (Loizeaux, 1999).

In addition to random factors, one of the difficulties in modeling progressive collapse caused by a bomb is order-of-magnitude differences in time scales between an explosive event and the response of the building. For example, the time frame in which damage occurs depends on the size of the explosion and the mechanism of resistance of the structural elements (floor slabs, walls, columns, etc.) and ranges from milliseconds to seconds. If the significant damage mechanism is the long-term dynamic response of the building, rather than just the explosion, millions of time steps will be required to simulate an explosion-induced collapse. By comparison, the time scale for earthquake-induced building response is on the same order as the seismic event that caused it. Consequently, modeling earthquake-induced damage will require far fewer time steps (Attaway, 1999).

Analytical models can still be useful for evaluating different designs, however, and, with continued refinements in hardware, software, and test data, they could be used to improve predictions of collapse mechanisms. The committee recommends that DTRA take advantage of the advances in parallel-processor computing made by DoD and the U.S. Department of Energy to improve the capability and ease of use of computational tools for predicting structural responses to bomb blasts. As the complexity of analytical models increases, verification and validation (V&V) of the resulting codes and solution techniques will be necessary. The experimental database being developed by the BMSP should be designed from the outset with the requirements for V&V in mind.

Small-Scale versus Full-Scale Testing

Reduced-scale testing offers an excellent, lower cost means of learning about failure modes and loading parameters and obtaining other data that can be used to test results of computational models. For these purposes, testing reduced-scale models and components is far more cost-effective than full-scale testing. However, reduced-scale testing also has limitations, such as scale effects (i.e., no model, no matter its size, can reproduce the prototype building exactly).

If the critical local response is well understood, reduced-scale tests can be very effective for determining behavioral phenomena of the system. For example, the flexural behavior of a complex slab or wall system subjected to out-of-plane loading can be reliably investigated with small-scale models (but not so small that the behavior of the component material is altered significantly). This cannot be done confidently, however, if the failure is, or would be, in shear. A very small specimen might have a very high shear strength leading to a flexural failure, even though a larger scale test would have been terminated by shear.

Scaling down connection details can be difficult, or even impossible. Reduced-scale tests of the response of floor slabs can yield much information, but gravity effects cannot be scaled conveniently. Therefore, it may not be possible to determine in a small-scale test if a floor slab would have collapsed onto the floor below and triggered a progressive collapse. Scale effects for buildings in the five-story range are probably not significant, but for buildings of ten or more stories, column sizes may be large enough to have a significant effect. An excellent discussion of the effects of scale on experimental results can be found in *Design of Model Test Program for a Buried Field Shelter* (Newmark & Associates, 1965). The committee believes that the BMSP should include a mix of full-scale and reduced-scale component testing.

From an analytical and modeling standpoint, reduced-scale tests are valuable because they are simpler in size, cost, and complexity of the test setup. However, one of the challenges in validating a numerical method is identifying which components can be modeled and validated independently; models of phenomena that involve complex effects cannot be validated by a single type of measurement, or even by a single test. For example, when modeling the deformation of a floor slab or column under heavy static load, the analyst can assess the accuracy of the model for both the concrete and the reinforcing steel. However, if something goes wrong with a fully coupled model of dynamic blast loads on a complicated structural geometry, it is difficult to determine which part of the model is incorrect.

The interpretation of test results requires critical judgement by the analyst. Questions about system behavior can be addressed with small-scale structures, provided the experimenter understands the behavior of details and is careful when comparing the results of calculations to tests. For example, in some instances, agreement may be more apparent than real because the experimental results in a particular application, or at that stage of loading, may be insensitive to the assumptions made in developing the theoretical model. Nevertheless, the data will be generally useful, and, because the cost of the small-scale experiments is relatively low, tests can be repeated to increase confidence in the results.

Problems that could be investigated at reduced scale include the effects of a floor slab failing and impacting lower floors, the effects of interior walls redirecting a blast wave to the floor or ceiling, and the effects of rebar splices on the strength of floor slabs. In a postattack analysis of the Alfred P. Murrah Federal Building in Oklahoma City, the authors speculated that confinement of the concrete would have increased the toughness of the columns (Corley et al., 1998). For this reason, the committee believes a series of tests of quarter-scale models would be useful for determining the effects of transverse reinforcement on column toughness. Confinement levels should be those currently used in spirally reinforced columns and in special moment frames.

Damaged Test Articles

In one test conducted as part of the BMSP, a column in CTS-1 was severely damaged and, by most definitions, failed. However, for various reasons, the CTS-1 structure did not collapse. Based on this event, and other experiences of committee members, the committee concluded that the residual strength of damaged columns and other components is not well understood but could be of great value for predicting the performance of buildings, their potential for progressive collapse, and for determining the stability of damaged buildings during search and rescue operations. For example, following earthquake-induced shear failures in columns, the columns sometimes remain standing and sometimes collapse, depending on the location of inclined cracks, the bending stiffness of longitudinal bars, and the location/size of the ties (Sozen, 1999). If the bars have sufficient lateral support to prevent buckling and to resist the dead load on the column along with whatever concrete remains, the building will not collapse. Because of limitations on what is known and the vagaries of construction, the results are very difficult to predict. The committee recommends that the BMSP undertake a series of column tests to cause various levels of damage and measure their residual strength. Full-scale columns representative of those in buildings of ten stories or more should also be tested.

Because there are so many possible combinations of rebar and column dimensions, the tests should be designed to bound the behavior of a representative design. Although building a database of responses of many column sizes and shapes would be valuable, a database that could be used to validate numerical models would be even more valuable in the long term. Once numerical methods have been shown to be capable of predicting the observed response, then numerical methods could be used to explore the residual strengths of damaged components.

Component Testing

The continuity of structural members in both steel and reinforced concrete buildings is highly dependent on reliable steel connections. Although a considerable amount of data on the performance of mechanical connectors under earthquake loads are available, little, if any, public information is available on their performance under blast loads. Consequently, a series of tests to evaluate how well common rebar splices and connections can function after being damaged by blasts would be useful. Data could be gathered, for example, on the behavior of large bars in mechanical splices, welded connections, lap splices, or Type 2 mechanical splices. Because steel connections, like welding details, cannot be accurately scaled down for reduced-scale testing, full-scale testing would be necessary for steel construction components.

The BMSP program has a unique opportunity to clarify relationships between earthquake-resistant building designs and blast-resistant building designs, especially on a component/subsystem basis. Although there have been discussions in the design community of the benefits of seismic designs to blast resistance, a recent paper comparing the design requirements showed that even a design that meets Seismic Zone 4 requirements may not result in a building resistant to blast effects (Ettouney et al., 1998). The key to improved performance appears to be in the type and location of structural detailing (Woodson and Krauthammer, 1998). Therefore, the committee believes that BMSP should plan to conduct a series of component tests to identify detailing methods for blast-resistant construction. The results could be used to inform the building code process and could ultimately be incorporated into design guides, such as *Building Code Requirements for Structural Concrete* (ACI, 1999). Comparative costs of alternative designs could also be developed to encourage the adoption of these techniques in commercial buildings.

Component testing would also be useful for validating high-fidelity numerical models. Although not all combinations of splices and joint types can be accounted for in a series of tests, examples of different joint types under different loading conditions could be used to develop benchmarks for testing models. Once confidence in sophisticated numerical models has been established, they could be used to calibrate the simple structural models used in the design process.

Next Class of Structures

The BMSP has focused on the testing and analysis of generic flat-slab structures because of their perceived vulnerability to blast damage and progressive collapse. As part of the committee's evaluation, the value of conducting full-scale testing on another generic structure type—possibly a steel frame building—was considered. For example, if an economical, blast-resistant steel frame design could be identified, it could have widespread commercial appeal because it could be erected and enclosed rapidly. However, although the generic building approach seems to be a reasonable experimental paradigm, the committee concluded that DTRA should not proceed with another controlled test structure for several reasons.

First, many of the typical structure types used in military construction domestically and overseas (e.g., multistory, load-bearing masonry; high-bay, long-span structures with load-bearing walls, such as gymnasiums and dining halls; preengineered metal buildings; and multistory wood structures), although simple structurally, are vulnerable to even fairly small bombs and usually have high occupancies. The observed performance of unreinforced masonry and other lightly constructed buildings, such as wood and preengineered metal, in earthquakes and high wind conditions has shown that they are very vulnerable to extreme loading. Thus, the potential for injury and death to a large number of occupants is probably higher in these buildings than in larger, more complex structures. On this basis alone, the BMSP would be justified in investigating these structures, and the committee recommends that DTRA consider including a series of component and reduced-scale tests on masonry structures in the BMSP. The test program should include unreinforced masonry for benchmarking purposes and a range of reinforcement techniques, including fully reinforced masonry, to improve protection. Following analysis and interpretation of the test results for both unreinforced and strengthened masonry, DTRA could conduct another series of tests on construction typical of long-span buildings.

The committee also questions the validity of the results obtained from repetitive blast loading of CTS-1. After the initial blast test, the structure became a *de facto* damaged structure. At that point, although the structure still has considerable value as a reaction frame for testing full-scale structural components and nonstructural elements, performance data for the overall building system obtained in subsequent tests will necessarily be compromised to some degree by preexisting damage. (For this reason, the committee strongly recommends that CTS-1 not be tested to failure.) The relative value of testing full-scale structures versus analysis, computational modeling, and component testing should also be reevaluated. From the standpoint of cost-effectiveness, the construction of another full-scale test structure is not justified at this time. Although a full-scale test structure could be valuable in a "proof-of-principle" demonstration, the committee believes that the next full-scale test structure should not be designed or constructed until a thorough study has been done to identify the knowledge gap(s) that should be addressed.

Internal versus External Blasts

The security community has expressed concerns about the vulnerability of interior columns and adjacent floor slabs to small satchel and suitcase bombs (~50 pounds) with little or no standoff (Dadazzio, 1999). Information on the damage potential of these small devices can be obtained from component and reduced-scale tests. Full-scale testing of a complete structure provides only a few data points, but with reduced-scale testing, repeated tests can be and should be run to reduce the considerable statistical variance in these events. In conjunction with good analytical models, reduced-scale testing (for reinforced-concrete columns) or full-scale component testing (for steel columns) can provide a basis for estimating the potential for structural collapse. By studying how joint construction affects the strength of components under explosive loads, it may be possible to create design guidelines that would lead to better methods for joining floor slabs and columns This information could then be used to estimate the potential for collapse.

In recent years, the American Concrete Institute committee (ACI-318) that deals with code requirements for structural concrete has added several requirements for detailing reinforcement to maintain building integrity. The committee believes the BMSP would benefit from obtaining data evaluating the reinforcement patterns of existing buildings and the reinforcement patterns now required by ACI-318, on the assumption that current ACI requirements have increased the resistance to progressive collapse following removal of a single column. If the BMSP tests prove otherwise, then other details to increase resistance could be evaluated. For example, some data are available on the resistance of slender elements to unusual loads over long spans, but investigations have not been made of catenary action with representative details.

As demonstrated by the World Trade Center bombing, a van filled with explosives in an enclosed parking area can do a great deal of damage. In these cases, the likely mechanism of collapse is very different than in external building explosions. Instead of loading an exterior column(s), an interior blast directly loads floor slabs, both above and below. Loss of the floor slabs destroys the lateral support of the remaining columns, further weakening the structure and rendering it prone to progressive collapse. Many parking garages are constructed of post-tensioned concrete elements (e.g., long-span, quad tees) that carry most of the load in continuous steel cables that may be very sensitive to progressive collapse if they are damaged by an explosion. Because of widespread concerns about vehicle bombs, the committee believes that DTRA should consider investigating parking garages as part of the BMSP.

Existing Buildings as Test Articles

The committee debated the value of using an existing building scheduled to be demolished as a test specimen. On the one hand, existing buildings may offer a technical advantage over purpose-built test structures because they are constructed by typical contractors following the plans and specifications of typical engineers using standards generally imposed at the time of the design. On the other hand, because of the inherent variability in construction details, they will be difficult to characterize from an analytical standpoint. Several problems would have to be addressed before buildings scheduled for demolition could be considered as viable experimental options. The most significant issue is that the building must be demolished within a few seconds after the test because of the very high probability that the structure will become unstable and pose a danger to the demolition crew. In addition, collecting data would require high-speed video, contact gauges, and displacement measurements—all before collapse occurred (a duration of seconds). Nevertheless, a purpose-built test structure at ERDC or White Sands, although easier to characterize, may be perceived as a specimen created especially for the test and, therefore, not representative, regardless of whether it was constructed in accordance with current procedures and standards.

Retrofits

Information about the design of new buildings can only be put into place as they are constructed. Because the inventory of existing buildings (many of which have vulnerabilities to blast loadings) far exceeds the number of new buildings, the committee believes that the BMSP should put a high priority on developing and evaluating retrofitting techniques—particularly creative conceptual retrofitting measures that would prevent a life-threatening progressive collapse following a blast that has destroyed the load-carrying capacity of a limited number of structural components. This recommendation is based partly on the relatively long service lives of buildings. For example, when the California Hospital Seismic Safety Act became law in 1972, hospitals already constructed were expected to be replaced by new facilities through attrition. However, as of 1990, more than 66 percent of California's hospitals were still not in compliance with the 1972 standard and were not be expected to be functional after a major earthquake (Office of Statewide Health Planning and Development, 1990).

Many retrofitting methods, including wrapping columns with carbon and glass fibers or jacketing them with steel plate and grout, are currently under consideration for inclusion in the BMSP. Methods for using carbon fiber as floor reinforcement against uplift have also been proposed. However, the absence of test data to support design calculations has seriously hampered the introduction of new, potentially life-saving technologies. Testing retrofitting methods will require that the behavior of the building without the retrofitting be determined first. The baseline behavior can then be used to assess the cost benefit of a given retrofitting design.

Testing and analysis of the reduced-scale structures at ERDC and the full-scale CTS-1 have demonstrated that some component behavior can be studied independently of overall structural behavior. A cost-effective way of adding a large amount of information to the design database would be to conduct a few carefully designed experiments to test a large number of retrofitting methods simultaneously. The BMSP's first priority should be evaluating available techniques, such as fiber wrapping and steel jacketing of columns, followed by techniques for improving the performance of floor slabs, particularly by strengthening them against uplift pressures. Technical analyses should be accompanied by economic data to aid engineers and other participants in the decision-making process. DTRA might consider partnering and cost-sharing arrangements with the private sector to accelerate the testing and deployment of emerging retrofitting technologies.

Multihazard Mitigation

The identification and assessment of design features and materials that can improve performance over a range of hazards (i.e., earthquake, fire, extreme wind events, chemical and biological agents) could have an ancillary benefit of improving building performance. Because design features that provide multihazard resistance are likely to generate more interest among designers and manufacturers than design features that promise only blast resistance, multihazard features could ultimately reduce the cost and increase the application of improved building practices and products.

The BMSP should investigate construction techniques that not only mitigate blast effects but also permit the rapid repair, recovery, and continued use of damaged buildings. Obviously, preventing progressive collapse is the primary goal, but if other building systems, such as electrical service and distribution, air conditioning, and fire and life-safety systems, can be restored promptly, the diminished loss of revenue might make mitigation cost effective for owners. Therefore, the committee believes that life-safety system elements should be included in the testing program of the BMSP. Components that could be tested include fire alarm control panels and sensors, fire-suppression components, and emergency lighting fixtures. At the same time, protective enclosures around critical building systems to reduce shock effects on electronic equipment and utilities could also be tested. If these are effective, they could both enhance the ruggedness of communications and prevent secondary fires and explosions from, for example, damaged gas valves. A coordinated program of testing and analysis could determine minimum distances between redundant building systems (e.g., backup and main power boards) and provide a technical and economic basis for providing redundant building systems. The results of the program could contribute not only to safer designs for high-hazard locations but could also support postattack rescue and recovery operations.

REDUCING INJURIES THROUGH NONSTRUCTURAL APPROACHES

Nonstructural Systems

The BMSP Master Plan states:

> The protection of occupants of buildings from terrorist bomb attacks can be enhanced by an appropriate balance between better security procedures including the enforcement of increased standoff distances and the use of blast mitigation techniques. This program addresses only the blast hardening and blast mitigation aspects of the problem including design of blast walls, blast loading modification, and structural hardening (DTRA, 1999).

In general, the BMSP has focused its efforts on building structures and related exterior wall components (i.e., structural approaches and physical design methods) for mitigating the blast effects, and consequently reducing the injuries and deaths, caused by terrorist bombs. Although the committee recognizes that the first task in designing a building to sustain a bomb blast is to prevent progressive collapse, once that has been achieved, minimizing loss of life and property from other blast effects becomes paramount. Substantial harm can be done even if structural collapse does not occur. For example, the board that reviewed the embassy bombings in Nairobi and Dar es Salaam in 1998 found:

> The damage to the embassy was massive, especially internally. Although there was little structural damage to the five story reinforced concrete building, the explosion reduced much of the interior to rubble—destroying windows, window frames, internal office partitions and other fixtures on the rear side of the building. The secondary fragmentation from flying glass, internal concrete block walls, furniture, and fixtures caused most of the embassy casualties. (U.S. Department of State, 1999).

Although evidence that window glazing is a major contributor to blast-related injuries and also causes many deaths is overwhelming (Gans and Kennedy, 1996; Mallonee et al., 1996), the potential of nonstructural internal building configurations and components to cause and mitigate injury is not well understood. Some attempt has been made to address this problem for earthquakes. For example, a report by the Federal Emergency Management Agency, *Identification and Reduction of Nonstructural Earthquake Hazards in Schools,* states:

> Nonstructural hazards can occur in every part of a building and include everything *but* the columns, beams, floors, load-bearing walls, and foundations. Common nonstructural items include ceilings, lights, windows, office equipment, computers, files, air conditioners, electrical equipment, furnishings, and anything stored on shelves or hung on walls. In an earthquake, nonstructural elements may become unhooked, dislodged, thrown about, and tipped over; this can cause injury and loss of life, extensive damage, and interruption of operations. (FEMA, 1993).

The Building and Fire Research Laboratory of the National Institute of Standards and Technology has reported that existing building codes do not adequately address the performance of nonstructural building components in earthquakes (Phan and Taylor, 1996). This report recommends research to assess the adequacy of current building codes and identify necessary improvements; to develop techniques to mitigate damage to ceiling components designed to older codes; and to develop uniform guidelines for the design, installation, and restraint methods for ceiling components.

Types of architectural partition and anchoring, interior building systems, and other nonstructural components can protect occupants by deflecting blast energy and shielding them from flying debris. At the same time, they also have the potential to become harmful debris themselves. The committee believes that the BMSP should focus more attention on the behavior of nonstructural systems in the blast environment, including tests of the effectiveness of various types of interior partitions or perimeter zones of "soft" space in protecting occupants and contents, as well as comparisons of floor-based systems of mechanical and electrical distribution and typical overhead systems.

The committee believes that CTS-1 can be used effectively for further testing of both structural and nonstructural elements. Outside protective features, such as perimeter walls of various heights and construction located at various standoff distances, berms, and other landscape features could also be tested with CTS-1 as the structure to be protected. These tests could identify combinations of building surface and internal features that would reduce blast effects to acceptable levels. The testing program should take into account that buildings are not just structures but interconnected series of systems that can work together to improve performance and increase safety.

Modeling for Injury Prediction

The modeling for injury prediction in the BMSP Program Plan appears to be based on a good understanding of the technical issues and tasks involved. However, the committee questions the quality of existing empirical data on human injuries caused by building failures and, thus, their suitability for use in epidemiological analyses. Both the quantity and quality of data used in analyses and validations will be critical to the development of reliable models. The epidemiologic and engineering literature on risk factors for physical injuries from natural and man-made disasters that involve building failures (e.g., earthquakes, tornadoes, hurricanes, volcanic explosions) is a potentially rich source of data and should be reviewed (e.g., Abrams et al., 1998; Jones et al., 1990; Tanaka et al., 1999). This literature includes data on injuries caused by sources other than structural components of buildings, including occupant behavior and damage to building contents and nonstructural components. The engineering literature, which is more extensive than the epidemiologic literature on injuries caused by building damage, should also be reviewed. However, because most engineers are not trained in epidemiologic methods designed to reduce bias in the interpretation of data, the BMSP should involve both epidemiologists and engineers in developing injury prediction models.

The committee notes that the BMSP's injury modeling does not include the so-called "crush syndrome," a class of serious and fatal injuries related to building damage and collapse that occurs when heavy objects, such as collapsed brick walls, pin individuals down. Crush-syndrome victims may suffocate rapidly or suffer life-threatening internal injuries even in the absence of blast-related trauma (Better, 1999). Most injury assessment tools do not include crush-syndrome casualties, but the committee believes crush syndrome should be included in the BMSP's efforts to model blast injuries.

Modeling and testing can contribute a great deal to an understanding of how to prevent injuries to the occupants of buildings. Postevent field studies are valuable tools for collecting information under less-than-ideal, real-world conditions. These types of studies have been done after both seismic events and bomb blasts (Durkin and Theil, 1992; Mallonee et al., 1996). A key aspect of collecting data is interviewing rescuers, victims, and bystanders (both injured and uninjured) to assess the factors that contributed to their situations and outcomes, including the rapidity, safety, and ease of rescue. However, access to these building occupants decreases rapidly over time, as do memories of the events. Therefore, the BMSP could provide a valuable tool by formalizing an institutionalized process that could be quickly engaged for data collection, analysis, and dissemination. The development of this process before an event occurs would accelerate the mobilization and deployment of trained investigative team(s), with prearranged funding and logistics, that could collect data on building damage, occupant injuries, and rescue difficulties in a standardized format. Injury data should not be limited to occupant injuries but should include injuries to rescuers as well (e.g., physical trauma and hazardous exposures [asbestos, chemicals, sewage, electrical wiring, etc.]). If these data were evaluated for many events by disaster epidemiologists, they could be valuable for continued improvements in design and construction, as well as rescue and recovery (Wagner et al., 1994). The Centers for Disease Control and Prevention, which fielded ad-hoc teams in the aftermath of the Oklahoma City and African embassy bombings, could be a valuable partner in an institutionalized program of injury reduction through improved design and rescue and recovery methods.

The bombing of the World Trade Center in 1993 demonstrated that fire and smoke propagation following a bomb explosion can be a significant cause of injuries to the building occupants (NRC, 1995). The flammability and smoke generation potential of building materials, furniture, and common office products have been studied extensively by many fire research organizations, including the Building and Fire Research Laboratory of the National Institute of Standards and Technology (BFRL). This work is included in FIREDOC, a web searchable bibliographic database of 55,000 holdings maintained by BFRL (BFRL, 2000). The committee believes that FIREDOC is a potentially rich source of data that can be useful for extending the range of injury prediction models to include injuries related to fire and smoke propagation.

Rescue and Recovery

Experience has shown that the most effective disaster planning is based on good evidence and data from past events (Auf der Heide, 1989). Critical data on blast injuries in buildings include:

- geographical patterns of injuries and their associations with design features, building materials, and building contents
- possibilities of escape from, or survival in, an attacked building
- accessibility to rescuers
- knowing where to look for survivors

Although damage-resistant building designs are critical in preventing injuries from terrorist attacks and other disasters, the ease and rapidity with which trapped or injured occupants can be extricated is also important. Key factors are how easily occupants can be located, whether they are able to evacuate the building, and whether rescuers can safely enter areas of the collapsed structure to render aid. The BMSP should evaluate these factors in cooperation with urban search and rescue teams and other rescue units to plan and carry out simulated rescue and recovery operations to refine or improve techniques.

INTERNATIONAL PROGRAMS

To date, the committee has heard only passing references to joint programs between DTRA and similar agencies in Israel and the United Kingdom. Although these programs are not funded by TSWG, the committee believes that similar work being carried out in other countries may offer DTRA opportunities to enhance program elements that are funded by TSWG, such as the BMSP. Some measure of international information sharing is currently being accomplished through the International Symposium on Interaction of the Effects of Munitions on Structures, a biennial event jointly sponsored by DTRA and the German Federal Ministry of Defence. The committee will review this activity, as well as other means for DTRA to incorporate the results of studies in other countries into its technology transfer efforts.

CONCLUSIONS AND RECOMMENDATIONS

Conclusion 2. Although the Program Master Plan includes many activities that could yield worthwhile benefits, the committee identified several modifications to the BMSP to be considered in the next program cycle.

Recommendation 2a. All analytical and experimental activities should be designed to test a specific hypothesis about the outcome. With respect to full-scale tests, parametric studies should be conducted to determine what could be learned from the test on the basis of the proposed instrumentation.

Recommendation 2b. The program should take full advantage of the advances in parallel-processor computing made by the U.S. Department of Defense and the U.S. Department of Energy to improve the capability and ease of use of computational tools for predicting structural responses to bomb blasts.

Recommendation 2c. The residual strength of blast-damaged structural components should be investigated more fully. For example, tests of full scale columns representative of buildings ten stories and more should be included, as well as a series of tests to evaluate how well common rebar splices and connections can function after being damaged by blasts.

Recommendation 2d. The Blast Mitigation for Structures Program should consider conducting a series of tests on masonry structures, including tests of unreinforced masonry for benchmarking purposes and tests of a range of reinforcement techniques to improve protection. A series of tests on construction typical of long-span buildings should also be considered.

Recommendation 2e. The Blast Mitigation for Structures Program should place a higher priority on the development and evaluation of retrofitting techniques—particularly on creative conceptual retrofitting measures that would prevent a life-threatening progressive collapse following a blast.

Recommendation 2f. The Blast Mitigation for Structures Program should focus more attention on the behavior of nonstructural systems in the blast environment, including (1) tests of the effectiveness of various types of interior partitions or perimeter zones of "soft" space in protecting occupants and contents, and (2) comparisons of floor-based systems of mechanical and electrical distribution and typical overhead systems.

Recommendation 2g. The Blast Mitigation for Structures Program should evaluate the key factors affecting the ease and rapidity with which trapped or injured occupants can be extricated from damaged buildings and whether rescuers can safely enter areas of the collapsed structure to render aid. In cooperation with urban search and rescue teams, the program should support simulated rescue and recovery operations to refine or improve rescue techniques.

Conclusion 3. The design and engineering approaches favored by the industrial contractors and government laboratories that are implementing the BMSP may be more appropriate to traditional military and defense objectives despite the emphasis of the BMSP on nonstructural solutions, injury reduction, and improved rescue and recovery techniques.

Recommendation 3. The contractor base should be broadened to increase the representation of the commercial architectural and engineering communities, as well as specialists in injury prevention, disaster medicine, and technology transfer, particularly in the planning phases of the program.

Conclusion 4. The committee is in complete agreement with the BMSP's emphasis on determining progressive-collapse vulnerability of buildings in selected attack scenarios but believes this ability would be improved by fuller coordination of research activities.

Recommendation 4. The Defense Threat Reduction Agency should adopt a balanced approach toward understanding and preventing the progressive collapse of buildings. This approach should include coordinated physical testing, experimentation, and analyses and should guide the planning of research activities and the interpretation and synthesis of the results.

Conclusion 5. Full-scale testing of structural systems has been overemphasized at this relatively early stage of the program at the expense of reduced-scale testing, the development of retrofitting techniques for existing buildings, the testing of nonstructural building systems, and the investigation of technologies related to injury prevention.

Recommendation 5. The Defense Threat Reduction Agency should not construct another full-scale test structure until the results of previous experiments on Controlled Test Structure-1 (CTS-1) have been fully analyzed and understood. At this stage of the program, DTRA should rely more on experiments with scaled elements and scaled assemblies of elements wherever scale effects are well understood.

Conclusion 6. Controlled Test Structure-1 has been underutilized so far; although it has been damaged in previous tests, it still has considerable value for testing full-scale structural components and nonstructural elements.

Recommendation 6. Controlled Test Structure-1 should not be tested to failure because it can still be used as a reaction frame for component tests.

Conclusion 7. Although, the inventory of existing buildings vulnerable to blast damage far exceeds the number of new buildings that will be constructed in the foreseeable future, the BMSP appears to have placed more emphasis on methods applicable to new construction than on retrofitting techniques for existing structures.

Recommendation 7. The development of tools for conducting vulnerability assessments and strengthening existing buildings should be given a higher priority. Resources should also be allocated to investigating construction techniques that permit the rapid rehabilitation of blast-damaged buildings.

Conclusion 8. The Blast Mitigation for Structures Program has a unique opportunity to determine how requirements and techniques for earthquake-resistant designs could apply to blast-resistant designs, as well as to identify and assess design features and materials that could improve building performance over a range of hazards (e.g.., earthquake, fire, flood, and extreme wind) that could impact the safety of the occupants.

Recommendation 8. The Blast Mitigation for Structures Program should incorporate activities with the maximum potential for multihazard mitigation. Because design features that provide multihazard resistance are likely to generate more interest among designers and manufacturers than design features that promise only blast resistance, multihazard features could ultimately reduce the cost and increase the application of improved building practices and products.

Conclusion 9. Data on blast-related injuries and building damage are limited and, therefore, have hindered the development of statistically valid damage-prediction and epidemiological models.

Recommendation 9. The Blast Mitigation for Structures Program should initiate an institutionalized process that can be quickly mobilized for collecting critical data related to blast damage and injuries in buildings that are subject to bomb damage.

REFERENCES

Abrams, J., L. Bourque, J. Kraus, C. Peek-Asa, D. Vimalachandra, and J. Yu. 1998. Fatal and hospitalized injuries resulting from the 1994 Northridge earthquake. International Journal of Epidemiology27: 459–465.

ACI (American Concrete Institute). 1999. Building Code Requirements for Structural Concrete and Commentary. Farmington Hills, Mich.: American Concrete Institute.

Attaway, S. 1999. Personal communication from Stephen Attaway, Distinguished Member of the Technical Staff, Sandia National Laboratories, Albuquerque, New Mexico, to Richard Little, director, Board on Infrastructure and the Constructed Environment, National Research Council, Washington, D.C., November 11, 1999.

Auf der Heide, E.1989. Disaster Response: Principles of Preparation and Coordination. St. Louis, Mo.: C.V. Mosby Company.

Better, O.S.1999. Rescue and salvage of casualties suffering from the crush syndrome after mass disasters. Military Medicine164(5): 366–369.

BFRL (Building and Fire Research Laboratory),2000. FIREDOC. Available on line at: *http://fris.nist.gov/cgi-bin/starfinder/0?path=firedoc. txt&id=anon&pas=anon&OK =OK*

Corley, W.G., P. Mlakar Sr., M. Sozen, and C. Thornton. 1998. The Oklahoma City bombing: summary and recommendations for multihazard mitigation. Journal of Performance of Constructed Facilities12(3): 100–112.

Dadazzio, R. 1999. Comment by Raymond Dadazzio, Weidlinger Associates, at the DTRA/TSWG Program Review on Structural Collapse, Alexandria, Virginia, November 4, 1999.

DTRA (Defense Threat Reduction Agency). 1999. Blast Mitigation for Structures Program Master Plan, June 1999. Alexandria, Va.: Defense Threat Reduction Agency.

Durkin, M.E., and C.C. Theil Jr. 1992. Improving measures to reduce earthquake casualties. Earthquake Spectra 7(1): 95–113.

Ettouney, M., R. Smilowitz, and R. Dadazzio. 1998. Comparison between Design Requirements of Earthquake and Blast Events. Paper T210-1 in *Structural Engineering World Wide 1998,* edited by N.K. Srivastava. Available on CD-ROM from Elsevier Science, New York.

FEMA (Federal Emergency Management Agency). 1993. Identification and Reduction of Nonstructural Earthquake Hazards in Schools. FEMA-241. Washington, D.C.: Federal Emergency Management Agency.

Gans, L., and T. Kennedy. 1996. Management of unique clinical entities in disaster medicine. Emergency Medicine Clinics of North America14(2): 301–326.

Jones, N.P., F. Krimgold, E. Noji, and G. Smith. 1990. Considerations in the epidemiology of earthquake injuries. Earthquake Spectra6(3): 507–528.

Loizeaux, M. 1999. Personal communication from Mark Loizeaux, CEO, Controlled Demolition Incorporated, Phoenix, Maryland, to Richard Little, director, Board on Infrastructure and the Constructed Environment, National Research Council, Washington, D.C., November 12, 1999.

Mallonee, S., S. Shariat, G. Stennies, R. Waxweiler, D. Hogan, F. Jordan. 1996.Physical injuries and fatalities resulting from the Oklahoma City bombing.Journal of the American Medical Association 5: 382–387.

Newmark and Associates. 1965. Design of Model Test Program for a Buried Field Shelter.Contract Report No. I-110. Report prepared for U.S. Army Engineer Waterways Experiment Station. Vicksburg, Miss.: U.S. Army Corps of Engineers.

NRC (National Research Council) 1995. Protecting Buildings from Bomb Damage: Transfer of Blast-Effects Mitigation Technologies from Military to Civilian Applications.Washington, D.C.: National Academy Press.

Office of Statewide Health Planning and Development.1990. A Recommended Program to Seismically Strengthen Pre-Hospital Act Hospital Facilities: A Response to Milestone 4, Initiative 1.2, "California at Risk," December 1990. Sacramento, Calif. Building Safety Board, Office of Statewide Health Planning and Development, Division of Facilities Development and Financing.

Phan, L.T., and A.W. Taylor, 1996. State of the Art Report on Seismic Design Requirements for Nonstructural Building Components. NISTIR 5857. Gaithersburg, Md.. U.S. Department of Commerce.

Sozen, M. 1999. Personal communication from Mete Sozen, Kettelhut Distinguished Professor of Structural Engineering, Purdue University, to Richard Little, director, Board on Infrastructure and the Constructed Environment, National Research Council, Washington, D.C., December 11, 1999.

Tanaka H., J. Oda, A. Iwai, Y. Kuwagata, T. Matsuoka, M. Takaoka, M. Kishi, F. Morimoto, K. Ishikawa, Y. Mizushima, Y. Nakata, H. Yamamura, A. Hiraide, T. Shimazu, and T. Yoshioka. 1999. Morbidity and mortality of hospitalized patients after the 1995 Hanshin-Awaji earthquake. American Journal of Emergency Medicine 17(2): 186–191.

U.S. Department of State. 1999. Report of the Accountability Review Boards on the Embassy Bombings in Nairobi and Dar es Salaam on August 7, 1998, January 1999. Available on line at: *http://www.state.gov/www/regions/africa/accountability report.html*

Wagner, R.M., N.P. Jones, and G.S. Smith. 1994. Risk factors for casualty in earthquakes: the application of epidemiologic principles to structural engineering. Structural Safety13: 177–200.

Woodson, S.C., and T. Krauthammer. 1998. Recent Developments in Blast-Resistant Structural Detailing. Paper T210-3 in *Structural Engineering World Wide 1998,* edited by N.K. Srivastava. Available on CD-ROM from Elsevier Science, New York.

4

Technology Transfer

BACKGROUND

One part of the committee's charge was to recommend a strategy for technology transfer for advances identified by the BMSP in engineering, architectural, and building practices, disaster preparedness and recovery, and emergency medical services. The need for information on blast-resistant design and improved dissemination of information cuts across disciplines and levels of technical expertise, and the field of earthquake engineering can provide many models that can be adapted to meet this need. This chapter describes a recommended strategy for the BMSP to develop an action plan; DTRA to form partnerships among federal agencies, the academic community, and professional organizations; and to begin the process of technology transfer. As a first step, a workshop on technology transfer for the mitigation of blast effects should be scheduled this year.

NEEDS OF THE ENGINEERING COMMUNITY

In light of the continuing focus on blast effects as a design consideration, engineers, who may not be specialists in blast engineering, are likely to be called upon increasingly to participate in the design process. These engineers will need better information than is currently available to enable them either to develop blast-resistant designs or to serve in an advisory capacity to building owners. Professional societies may consider certifying engineers in blast-resistant design, much as they do for engineers specializing in earthquake-resistant design. Necessary information will include generic blast loads for various charge weights/shapes and standoff distances, structural design procedures, and detailing requirements. Although much of this information is already available in various manuals and technical documents produced by the military and other organizations (e.g., *Design and Analysis of Hardened Structures to Conventional Weapons Effects* [U.S. Army et al., 1997], *Structures to Resist the Effects of Accidental Explosions* [U.S. Army, 1990], and *Design of Structures to Resist Nuclear Weapons Effects* [ASCE, 1985]), its distribution is generally limited to government agencies and their contractors, and it can not always be easily applied to civilian structures. These manuals do provide an excellent starting point, however, for generic guidance on blast-resistant design. The military has attempted to address conventional construction in *Estimating Damage to Structures from Terrorist Bombs, Field Operations Guide* (USACE, 1999), but this is more of a vulnerability assessment than a design tool. The recently released report By the American Society of Civil Engineers (ASCE), *Structural Design for Physical Security: State of the Practice* (ASCE, 1999), is another important source of technical information. Related guidance documents include *Minimum Design Loads for Buildings and Other Structures* (ASCE, 1996) and *Recommended Lateral Force Requirements and Commentary* (SEAOC, 1996), both of which describe seismic loads and design approaches (the former addresses dead and live loads, as well as soil, flood, wind, snow, rain, ice, and earthquake loads, singly and in combination) and could serve as models for transferring current knowledge and advances in blast-resistant design as they become available. Structural systems to resist the forces of high winds and tornadoes are discussed at length in *Minimum Design Loads for Buildings and Other Structures,* (ASCE, 1996). Although the potential benefits of main wind-force resisting systems, such as cross-bracing and shear walls, for improving blast resistance have not received as much attention as seismic improvements, the wind-engineering community could provide valuable insights into improved multihazard building performance.

Blast-related information for nonengineers would appeal to a wide audience. This information could clear up some of the confusion about the effects of explosions and blast effects on structures, the benefits of setbacks, the cost and effectiveness of mitigation measures, and so on. A series of briefing papers published by the Applied Technology Council (ATC) and the Structural Engineers Association of California (SEAOC) on seismic design and construction (ATC and SEAOC, 1998) could serve as a model for a similar series on blast-resistant design published as part of the BMSP.

BUILDING CODES AND STANDARDS

The committee believes that provisions for blast-resistant design are not likely to be incorporated into model building codes in the foreseeable future. Although bombing attacks have serious consequences, the probability that a civilian building will be the target of a terrorist bombing is relatively low. Therefore, public support for blast-resistance requirements in all construction is also low, and building owners are reluctant to pay the additional costs of designing and maintaining blast-resistant features. The committee believes that a general market demand or some economic motivation, such as reduced insurance rates, will be necessary before commercial developers will act voluntarily. Without strong financial or regulatory incentives, blast-resistant design features are not likely to become common practice unless they provide more readily acceptable (and understandable) multihazard mitigation benefits. Damage-resistant/injury-preventive designs that can be effective against multiple hazards, such as earthquakes, fire, wind, and floods (which are much more likely to occur than bomb blasts), will be easier to promote and justify for a wider potential market. A larger market would, in turn, make the manufacture and distribution of materials and technology to meet these design specifications more economically attractive.

In light of the lack of public awareness of the nature of the threat and the lack of a consensus on what to do about it, guidance on building collapse and building code requirements would be valuable. The phenomenom of progressive collapse is not well understood, despite the widespread recognition that most catastrophic building failures involve building collapse. Although building codes admonish designers to ensure that collapses do not occur, they do not provide guidance on how this can or should be accomplished. Because preventing collapse is a key factor in reducing fatalities in the event of a terrorist bombing, suggested design approaches to reduce the likelihood of progressive collapse would serve the BMSP's purposes and be of immense value to the structural engineering community. Guidelines for applying building codes would also be helpful. *Building Code Requirements for Structural Concrete and Commentary* (ACI, 1999) devotes an entire chapter to special provisions for seismic design, including the detailing of reinforcing steel in concrete construction and the applicability of different structural frames. This document could serve as a vehicle for conveying similar provisions for blast-resistant designs that have been verified through the BMSP. However, resistance to dynamic loads of the three moment frames specified in *Building Code Requirements for Structural Concrete and Commentary* (i.e., normal, intermediate, and special) differs markedly. Therefore, building frames selected without taking dynamic forces imposed by natural hazards (e.g., earthquakes or extreme winds) into account will also be less likely to resist blast loads. Therefore, in addition to addressing specific code provisions, for the benefit of those seeking guidance in blast-resistant design, the BMSP's technology transfer strategy should also describe when and how the provisions should be applied.

The International Building Code has replaced the three model building codes followed in the United States, (i.e., the Uniform Building Code, Building Officials and Code Administrators National Code, Standard Building Codes). Although a single building code would appear to make it easier for the BMSP to evaluate commercial building codes, the three codes will remain in force in many jurisdictions for several years. Therefore, additional evaluations may be required. In addition to the commercial building codes, the BMSP should also be aware of several other documents that address dynamic effects. For example, Japan and New Zealand have developed extensive codes, procedures, and construction methods for structures subjected to earthquake motions. Furthermore, the design of mechanical, electrical, and other elements, especially those relating to physical security, fire, and life safety must also be addressed. Therefore, the BMSP should evaluate mechanical, electrical, and codes by other groups, such as the National Fire Protection Association, the Nuclear Regulatory Commission, the American Society of Mechanical Engineers, and the Institute of Electrical and Electronic Engineers. New editions of these documents are released every few years, and the design community must be kept aware of the continuous evolution of this body of knowledge. Revisions to the documents include both new requirements and changes to existing methods, based on evaluations of ongoing research.

The earthquake engineering field provides an excellent example of this evolutionary process. Beginning in 1959, seismic risk and design requirements were based on the *Blue Book* published by the Structural Engineers Association of California (SEAOC, 1996). The Applied Technology Council (ATC), which was established after the Sylmar earthquake in 1971, published *Tentative Provisions for the Development of Seismic Regulations for Buildings* in 1978 (ATC, 1982), under the sponsorship of the National Science Foundation (NSF) and the National Bureau of Standards (now the National Institute of Standards and Technology [NIST]). The promulgation of seismic design requirements was taken over by the Federal Emergency Management Agency (FEMA), which in 1997 issued the *National Earthquake Hazards Reduction Program Guidelines for Seismic Rehabilitation of Buildings and Commentary* (FEMA, 1997a) and *National Earthquake Hazards Reduction Program Recommended Provisions for Seismic Regulations for New Buildings and Other Structures and Commentary* (FEMA, 1997b). Recent versions of the Uniform Building Code have included recommendations from the *Blue Book* (as revised), as well as work sponsored by NIST and FEMA and performed by ATC, the Building Seismic Safety Council (BSSC), and other organizations. The array of publications and activities in the earthquake field have been facilitated by the Earthquake Engineering Research Institute (EERI) and the earthquake engineering research centers sponsored by the NSF, under the auspices of the National Earthquake Hazard Reduction Program (NEHRP).

ROLE OF ACADEMIA

The BMSP, a DoD research and development program, is not intended to support basic research. Nevertheless, the nature and mechanism of progressive collapse is a topic that merits further study by the academic community. This subject, which has not been addressed at a meaningful level for almost a quarter of a century, could have benefits for mitigating risks from many hazards. Coordinating basic research with the BMSP and ensuring a funding source will require partnerships (e.g., between NSF, the BFRL, DoD, and DTRA). Similar partnerships by NEHRP to leverage funds for earthquake research could provide a model of interagency cooperation and a model for securing congressional support for basic research to address national-level issues.

Since the early 1980s, active involvement of academicians in fortification-related research has declined as the result of attrition and reduced research support. The NRC report, *Protecting Buildings From Bomb Damage: Transfer of Blast-Effects Mitigation Technology from Military to Civilian Applications* states:

> The committee has found that there are several serious barriers to technology transfer from the military to the civilian sector. The first major barrier is education. The current academic and professional training of architects and engineers does not adequately prepare the design professionals, either technically or philosophically, to incorporate blast-hardening principles in civilian structures. Thus, a strong educational commitment is required by university schools of architecture, construction, and engineering, as well as by professional engineering societies, if the potential for technology transfer is to be realized (NRC, 1995).

Universities could provide a significant contribution to technology transfer and closing the training gap by including aspects of blast-resistant design in their structures and structural dynamics curricula, similar to university programs in earthquake-resistant or wind-resistant design. Furthermore, training should be complemented by university involvement in active research related to the improved blast-resistance of structures.

University students could be directly involved in blast-related topics through internships with federal agencies, student design competitions for blast-resistant structures, and other mechanisms. Academic involvement in physical and computational research in this area will have the benefit of the focused research that typically occurs in the university environment. The role of academia could also be extended to market research and risk communication related to the promotion of hazard-resistant building design.

HANDLING OF SENSITIVE INFORMATION

A technology transfer program for mitigating blast effects must be tailored to handle potentially sensitive information. Although the design basis (underlying assumptions of design blast loads and the location and configuration of critical services) for a specific building would be valuable to terrorists, the process used to design the building would not. Plans for buildings, both public and private, which are often available from local building officials to anyone who requests them, could represent a more serious security issue than the widespread dissemination of design guidance. Nevertheless, the committee recognizes that the dissemination of test and analysis data for specific components from the BMSP coupled with detailed structural plans could be a serious security risk. However, realization of the benefits of a wide range of improved techniques, materials, and practices developed by the BMSP will require that the information not be restricted to a narrow group of users.

This issue has been addressed by the Blast Mitigation Action Group (BMAG) organized by the U.S. Army Corps of Engineers ERDC to identify and evaluate COTS products. The BMAG web site is password-protected and provides sources of blast-mitigation products and services for both retrofitting and original construction to registered users (BMAG, 2000). The committee supports this excellent initiative and believes that it should be expanded to disseminate information about a broader range of products.

A FRAMEWORK FOR TECHNOLOGY TRANSFER

The effective transfer of the results of the BMSP will require overcoming many barriers. One way to accelerate this process is to make maximum use of existing institutional infrastructures for disseminating information. These include joint activities with other organizations, such as ACI (especially Committee 370, Short Duration Dynamic and Vibratory Load Effects), the American Institute of Steel Construction (which is considering establishing a committee to address blast effects), and the American Society of Civil Engineers (which has established a committee to develop a standard for blast protection of buildings), to develop model code and standard provisions that would be voluntary in the civil sector but possibly mandatory for select government buildings. Although code- and standard-writing processes are lengthy and involved, the BMSP could provide data and information as it becomes available for possible inclusion in voluntary guidance documents (e.g., detailing methods for reinforcing steel to resist localized and progressive collapse). The participation of the American Institute of Architects will be critical in joint activities; the members of this highly respected organization are in a position to encourage the incorporation of blast-resistant features and designs.

The committee believes that the next step for the BMSP should be to convene a formal workshop in 2000 to document information needs and develop a plan of action for technology transfer in the mitigation of blast effects. The BMSP should also increase its efforts to provide continuing outreach to other federal agencies. An organization of the National Research Council, the Federal Facilities Council, was established to address information needs related to the planning, design, construction, operations, maintenance, and management of federal facilities, and could serve as an information portal to federal agencies.

The BMSP should also consider sponsoring an annual or biennial conference on blast-mitigation design and engineering. Although these issues are discussed at existing engineering, construction, security, and emergency management conferences, a single, integrated forum could be of enormous benefit in the dissemination of the latest advances in the field and could stimulate the development of new and effective retrofitting concepts for existing structures.

Figure 4-1 illustrates the committee's suggested model for a technology transfer strategy for the BMSP.

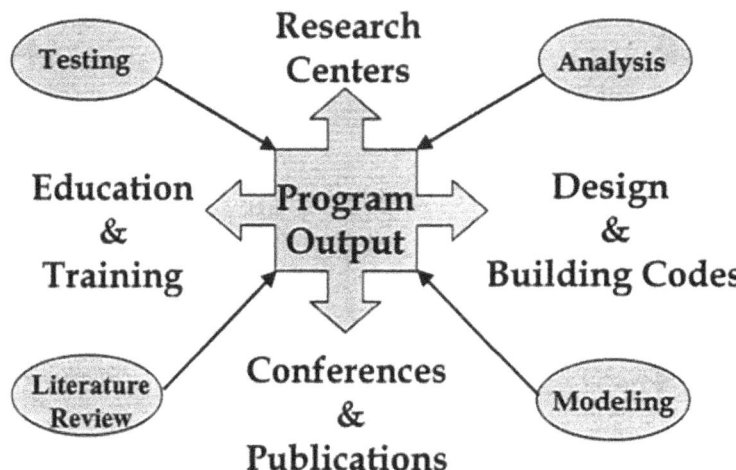

FIGURE 4-1 Model of a technology transfer strategy.

CONCLUSION AND RECOMMENDATION

Conclusion 10. The barriers to the complete and effective transfer of the results of the BMSP will require considerable time and effort to overcome. A convenient way to reduce the transfer time would be to use existing institutional infrastructures (i.e., building code and standards-writing organizations, professional and technical organizations, universities, and research centers) to disseminate knowledge.

Recommendation 10. A workshop to develop a road map for transferring technology for mitigating blast effects should be scheduled as soon as possible. To assist in the ongoing dissemination of information, the Blast Mitigation for Structures Program should consider sponsoring an annual or biennial conference devoted to all aspects of blast-mitigation design, engineering, injury prevention, and rescue and recovery.

REFERENCES

ACI (American Concrete Institute). 1999. Building Code Requirements for Structural Concrete and Commentary. Farmington Hills, Mich.: American Concrete Institute.
ASCE (American Society of Civil Engineers). 1985. Design of Structures to Resist Nuclear Weapons Effects. New York: American Society of Civil Engineers.
ASCE. 1996. Minimum Design Loads for Buildings and Other Structures. Reston, Va.: American Society of Civil Engineers.
ASCE. 1999. Structural Design for Physical Security: State of the Practice. Reston, Va.: American Society of Civil Engineers.

ASCE. 1999. Structural Design for Physical Security: State of the Practice. Reston, Va.: American Society of Civil Engineers.

ATC (Applied Technology Council). 1982. Tentative Provisions for the Development of Seismic Regulations for Buildings (ATC-3-06). Redwood City, Calif.: Applied Technology Council.

ATC and SEAOC (Applied Technology Council and Structural Engineers Association of California). 1998. Built to Resist Earthquakes: The Path to Quality Seismic Design and Construction. Redwood City, Calif.: Applied Technology Council.

BMAG (Blast Mitigation Action Group). 2000. BMAG. Available on line at:*http://bmag.nwo.usace.army.mil*

FEMA (Federal Emergency Management Agency). 1997a. National Earthquake Hazards Reduction Program Guidelines for Seismic Rehabilitation of Buildings and Commentary. Washington, D.C.: Federal Emergency Management Agency.

FEMA. 1997b. National Earthquake Hazards Reduction Program Recommended Provisions for Seismic Regulations for New Buildings and Other Structures and Commentary. Washington, D.C.: Federal Emergency Management Agency.

NRC (National Research Council). 1995. Protecting Buildings from Bomb Damage: Transfer of Blast-Effects Mitigation Technologies from Military to Civilian Applications. Washington, D.C.: National Academy Press.

SEAOC (Structural Engineers Association of California). 1996. Recommended Lateral Force Requirements and Commentary. Sacramento, Calif.: Structural Engineers Association of California.

USACE (U.S. Army Corps of Engineers). 1999. Estimating Damage to Structures from Terrorist Bombs, Field Operations Guide. Engineering Technical Letter 1110-3-495. Washington, D.C.: U.S. Army Corps of Engineers.

U.S. Army. 1990. Structures to Resist the Effects of Accidental Explosions. TM 5-1300. Washington, D.C.: Department of the Army.

U.S. Army, U.S. Air Force, U.S. Navy, and Defense Special Weapons Agency. 1997. Design and Analysis of Hardened Structures to Conventional Weapons Effects. TM 5-855-1/AFPAM 32-1147(I)/NAVFAC P-1080/DAHSCWEMAN-97. Washington, D.C.: U.S. Army Corps of Engineers.

Appendixes

APPENDIXES

A
Biographies of Committee Members

Mete A. Sozen (chair), NAE, is Kettelhut Distinguished Professor of Structural Engineering, at Purdue University. Dr. Sozen specializes in teaching and research related to the analysis and design of concrete structures subject to earthquake and other dynamic loadings. He is the recipient of many awards and honors including: Lindau Award and Kelly Award, American Concrete Institute (ACI); Boase Award, Reinforced Concrete Council; Research Prize, Howard Award and R.C. Reese Award, American Society of Civil Engineers (ASCE); General Electric Senior Research Award of the ASCE, Lifetime Achievement Award of The Illinois Section of ASCE, the Parlar Science Award of Middle East Technical University (Turkey); and election to the Royal Swedish Academy of Engineering Sciences. He is an honorary member of the Association of Turkish Engineers, ASCE, and the ACI. He has served as a member of the Veterans Administration Advisory Committee on Structural Safety; ACI Technical Activities Committee; ACI Committee 318 on the Building Code; and chair of the United States-Japan Cooperative Research Program on Urban Earthquake Hazard Reduction. In 1996, he participated in the Federal Emergency Management Agency's Building Performance Assessment Team's investigation of the Oklahoma City bombing. Dr. Sozen has served as the chair of the National Research Council Committee on Natural Disasters and Committee on Earthquake Engineering. He holds honorary doctorates from Bogazici University (Turkey) and Janusz Pannonious University (Hungary) and has earned a B.S. in civil engineering from Robert College, and an M.S. and Ph.D. in civil engineering from the University of Illinois at Champaign-Urbana.

Stephen W. Attaway is a distinguished member of the Technical Staff at Sandia National Laboratories. His expertise is in computational simulation and modeling and code development for applications in solid mechanics. Dr. Attaway is currently leading research and development for the development of solid mechanics code at Sandia's Engineering Science Center, including the development of coupled codes for blast-structure interaction. He received his B.S., M.S., and Ph.D. degrees in civil engineering, all from the Georgia Institute of Technology.

Erik Auf der Heide is a medical officer with the Agency for Toxic Substances and Disease Registry of the U.S. Department of Health and Human Services. He is also an adjunct assistant professor in emergency medicine at the Emory University School of Medicine, where he completed a fellowship in disaster medicine. He is the author of two textbooks and a number of articles on disaster management. Dr. Auf der Heide is a Member of the Disaster Section of the American College of Emergency Physicians. He received his M.D. from Baylor College of Medicine, an M.P.H. in epidemiology from the Rollins School of Public Health, Emory University School of Medicine, and is certified by the American Board of Emergency Medicine.

W. Gene Corley, NAE is vice president of Construction Technology Laboratories. His expertise is in structural concrete for earthquake-resistant construction, the uses of concrete in buildings and bridges, investigations of structural failure, and building codes. Dr. Corley serves on the Technical Council on Forensic Engineering of the American Society of Civil Engineers and was the principal investigator on the Building Performance Assessment Team dispatched by the Federal Emergency Management Agency to study the Alfred P. Murrah Federal Building in the aftermath of the bombing. He received his B.S. and M.S. in civil engineering and his Ph.D. in structural engineering from the University of Illinois at Champaign-Urbana.

Eve Hinman, principal of Hinman Consulting Engineers, has experience in the design and analysis of structures subject to the effects of explosions and project experience in the effects of conventional and nuclear weapons, accidental explosions, and terrorist attack. She has provided consultant services to the General Services Administration, the U.S. Department of State, and the Port Authority of New York and New Jersey. Dr. Hinman has conducted many investigations of accidental explosions and the bombing of the Alfred P. Murrah Federal Building in Oklahoma City. She has a B.S. in civil engineering, an M.S. in structural engineering, and a D.Sc. in engineering mechanics, all from Columbia University.

Robert P. Kennedy, NAE is a consulting engineer in structural mechanics with expertise in structural dynamics, structures, earthquake engineering, engineering mechanics, design codes, and standards. He has experience in static and dynamic analysis and the design of special-purpose civil and mechanical-type structures, particularly for the nuclear, petroleum, and defense industries. He has designed structures to resist extreme loadings, including seismic landings, missile impact, extreme wind, impulsive loads, and nuclear environmental effects, and has developed computerized structural analysis methods. Dr. Kennedy received his B.S. in civil engineering and his M.S. and Ph.D. in structural engineering from Stanford University.

Sam A. Kiger is chairman and C.W. La Pierre Professor of Civil and Environmental Engineering and director of the National Center for Explosion Resistant Design (NCERD), at the University of Missouri, Columbia. His areas of expertise are structural analysis and design, structural dynamics, explosion effects, protective construction, and soil-structure interaction. The NCERD is an interdisciplinary center that serves as a national focus for university-based research to create new knowledge and improve understanding of the explosion environment and blast-mitigation technology; create new and improved structural designs and strategies for protection from explosions; and transfer technology through publications, short courses, and university degree programs. Prior to teaching, Dr. Kiger spent many years on the staff of the Waterways Experiment Station of the U.S. Army Corps of Engineers, where much of his work was focused on the effects of explosions on structures. He received his B.S., M.S., and Ph.D. in theoretical and applied mechanics from the University of Illinois at Champaign-Urbana.

Stuart L. Knoop is a registered architect, a fellow of the American Institute of Architects, and co-founder of Oudens and Knoop, Architects, PC, of Chevy Chase Maryland. He has been involved in design for security for many years, particularly for the U.S. State Department, Office of Foreign Buildings Operations. Oudens and Knoop has designed security upgrades for more than 60 embassies and consulates worldwide. Mr. Knoop has served on the National Research Council (NRC) Committee on Research for the Security of Future U.S. Embassy Buildings and was vice chair of the committee that produced the NRC report, *Protecting Buildings from Bomb Damage* (National Academy Press, 1995). He is a member of the NRC Commission on

Engineering and Technical Systems. Mr. Knoop holds a B.Arch. from Carnegie Institute of Technology.

Johanna LaPierre, associate vice president at RTKL Associates, Inc., has experience in architectural, interior, and landscape architecture on projects ranging from large-scale office developments and renovations to embassies, hotels, and historic restorations. Her responsibilities have included project management and coordination, design, production of contract documents, and contract administration. She has also been project manager for security upgrades at the U.S. Capitol, U.S. Supreme Court, and Library of Congress, as well as numerous U.S. Department of State facilities worldwide. She holds a B.A. from Cornell College and an M.Arch. from the University of Virginia.

Mark Loizeaux is chairman of the Loizeaux Group of Companies, chief executive officer of Loizeaux Group International, and president of Controlled Demolition, Inc. Mr. Loizeaux has 35 years of experience in both the conventional and explosives demolition trades and provides consulting services on demolition and site clearance, rock removal and vibration control, antiterrorist measures to mitigate the effect of attacks on structures, and forensic analysis of damaged structures to determine the cause of damage. He has been personally responsible for field supervision of the demolition of more than 1,200 structures worldwide and is internationally recognized as a leader in the demolition and explosives industry. Mr. Loizeaux holds a B.S. in business administration from the University of Tennessee.

J.L. Merritt, an independent consulting engineer in civil, geotechnical, and structural engineering, has extensive research and practical experience in the design of protective structures, soil-structure and blast-structure interactions, and earthquake engineering. He has published more than 70 articles, papers, and monographs. Prior to his consulting career he taught at the University of Illinois, where he attained the rank of professor. He is a member of numerous professional societies and a registered engineer in various specialties (civil, geotechnical, and structural) in four states. Dr. Merritt received a B.S. in civil engineering from Lehigh University and an M.S. in civil engineering and Ph.D. in engineering from the University of Illinois at Champaign-Urbana.

David J. Pelgrim, an engineer with E. K. Fox & Associates Consulting Engineers has been responsible for the design of numerous physical and technical security upgrade projects, the production of feasibility studies, the performance of field investigations, and the creation of construction documents. He has undertaken projects for numerous clients, including the U.S. Army Corps of Engineers, the U.S. Navy, the U.S. Air Force, the General Services Administration, the U.S. Department of State, the U.S. Department of Justice, and the District of Columbia. He has completed the design of systems at numerous facilities throughout the world, spanning all physical and technical threat levels. In addition to his work in facility security, Mr. Pelgrim has been engaged in a broad range of projects involving the design of power, lighting, fire protection, life safety, communications, automated controls, and other building systems. He has also been involved in the design of innovative building systems as a joint venture between the General Services Administration's Centers for Expertise and Carnegie Mellon University. Mr. Pelgrim holds a B.S. in mechanical engineering from the University of Maryland at College Park and is a registered professional engineer.

Eugene Sevin, NAE, is an independent consultant. His research interests are nuclear and conventional weapons effects, hardened facility design, and computational structural mechanics. He formerly served with the U.S. Department of Defense (DoD) as deputy director, Space and Missiles Systems, and with the Defense Nuclear Agency as assistant to the deputy director (Science and Technology) for experimental research. Dr. Sevin was professor of mechanical engineering at the Technion, Israel Institute of Technology, and head, Mechanical Engineering Department, at Ben Gurion University of the Negev, Israel. He was also adjunct professor of applied mechanics at the Illinois Institute of Technology (IIT) and director of engineering mechanics research at IIT's Research Institute. Dr. Sevin chaired the committee that produced the National Research Council report, *Protecting Buildings from Bomb Damage* (National Academy Press, 1995). He recently served on a peer review group for the U.S. Army Corps of Engineers, Waterways Experiment Station, and a Defense Science Board Task Force on Underground Facilities. Dr. Sevin has published extensively and has received a number of awards; in October 1998 he was the inaugural recipient of the DoD Shock and Vibration Information Analysis Center's Melvin L. Baron Award in structural dynamics and constitutive modeling. He earned a B.S. in mechanical engineering from IIT, an M.S. from the California Institute of Technology, and a Ph.D. in applied mechanics from IIT. He is a member of the American Society of Mechanical Engineers and the American Institute of Aeronautics and Astronautics and has served on numerous DoD and interagency committees.

Charles H. Thornton, NAE, is chairman of Thornton-Tomasetti Engineers/ the LZA Group, Inc., a 350-person organization that provides structural engineering and architectural services, failure analysis, hazard mitigation, and disaster response services. Dr. Thornton has provided expert witness testimony for many clients and is a recognized expert on collapse and structural failure analysis. He led the engineering investigation of the causes of the collapse of the Hartford Coliseum Space Truss Roof, the scaffold collapse at Pleasant Point Power Station, West Virginia, and the collapse of the New York State Thruway Schoharie Bridge. In 1996, he participated in the Federal Emergency Management Agency's Building Performance Assessment Team's investigation of the Oklahoma City bombing. Dr. Thornton is presently a visiting faculty member at Princeton University and Manhattan College; he has taught at Pratt Institute and Cooper Union. Dr. Thornton is a member of the Board of Trustees of Manhattan College, the Applied Technology Council, and the Building Seismic Safety Council. He holds a B.S. from Manhattan College and an M.S. and Ph.D. from New York University.

B

Blast Mitigation for Structures Program Master Plan

Draft In-Progress

Technical Support Working Group Physical Security Subgroup
TASK 148F
BLAST MITIGATION FOR STRUCTURES PROGRAM MASTER PLAN
June 1999

TASK MANAGER:
Doug Sunshine
Defense Threat Reduction Agency
6801 Telegraph Road
Alexandria, VA 22310

EXECUTIVE SUMMARY

The purpose of the Blast Mitigation Program is to develop methods to protect people inside of buildings from terrorist bomb attacks. Although this will be accomplished, in part, by structural hardening and blast mitigation techniques, the focus is on protecting people, not the building or other assets. Two key issues are addressed in the program which are the primary cause of injuries and death in terrorist bombings; structural collapse, and flying debris (i.e., glass, building materials, etc.). Products of the program include better assessment methods which identify potential vulnerabilities, and new methods and guidance on how to reduce these vulnerabilities. The program will develop methods to design new, less vulnerable structures, as well as develop methods to retrofit existing structures to improve their protection levels.

During the first year (FY98), the program has initiated the development of methods to retrofit existing structures to prevent structural collapse. A four story test building has been constructed at White Sands Missile Test Range (WSMR) to examine collapse mechanisms and test retrofit methods. Retrofits using traditional materials, such as concrete and steel, as well as the use of advanced composite materials will be tested. This building has also been used to test mailrooms to protect against package bombs, windows and window retrofit methods, and will hold various walls for additional blast testing. To address roof and load-bearing wall collapse, a reaction structure has been constructed to test retrofit concepts. Tests during FY98 provided baseline structural response information; retrofits will be tested during FY99. Out-year efforts will develop design guidance for retrofitting steel framed structures and other structural types. To address the design of new buildings, in cooperation with GSA, a design methodology to prevent progressive collapse will be developed and tested. Ultimately, guidelines will be produced for the design of all GSA buildings. These guidelines could be used by DoD and other government agencies.

Flying debris, particularly glass, is a major source of injury from terrorist bombings. Methods to predict the response of glass to blast have been limited to estimating glass breakage. The United Kingdom (UK) has developed a Glazing Hazard Guide which defines potential hazard levels from break-safe (little hazard) to high hazard. This guide was computerized as a first step in developing a comprehensive method to predict the hazards and injuries from flying glass. During FY98-99, in cooperation with the UK, ninety-six (96) windows were tested in the building at WSMR to verify the hazard guide at high explosive yields where little data exists. Future efforts will focus on more comprehensive models which predict fragment size, shape, velocity, and distribution from all window types and sizes. This model will be linked to human injury models to estimate hazards and compare the effectiveness of various mitigation techniques.

There are many commercial companies with various products which may help reduce injuries from glass. Unfortunately, there have been very few tests conducted in a consistent manner to evaluate the capabilities of these products. Testing was initiated during FY98 on sixty (60) products to compare their blast performance. The results will be published on a "Yellow Pages" web page which will help building owners understand the capabilities of, and differences between products. Testing of door and wall products will be conducted during FY00.

Validation tests were successfully conducted in Israel to verify design methods for masonry wall and window retrofits that provide inexpensive systems to reduce debris hazards. The wall system uses fabrics which do not strengthen the wall, but catch the debris thus protecting people inside. The window system uses fragment retention film attached to the glass and a bar across the window which, in combination, catches the glass. Design and implementation guidance has been completed and will be distributed. Additional retrofit methods will be tested during FY00-01.

Tests were conducted in Israel and WSMR on various blast wall designs. New, less expensive, and more aesthetically pleasing designs will be developed during FY99. Tests were also conducted during FY99 to validate blast wall shielding effects.

ACRONYMS AND ABBREVIATIONS

ASCE	American Society of Civil Engineers
ATF	Bureau of Alcohol, Tobacco, and Firearms
BAA	Broad Area Announcement
CMU	Concrete Masonry Unit
CONWEP	PC code for conventional weapons effects based on TM 5-855-1
CTS1	Full-scale test structure to be built and tested in FY98
CTS2	Full-scale test structure to be built in FY98 and tested in FY99
DoD	Department of Defense
DOE	Department of Energy
DOS	Department of State
DSWA	Defense Special Weapons Agency
FBI	Federal Bureau of Investigation
FCDSWA or FC	Field Command, Defense Special Weapons Agency
FEMA	Federal Emergency Management Agency
FY	Fiscal Year
GSA	General Services Administration
IACRO	Interagency Cost Reimbursement Order
MIPR	Military Interdepartmental Purchase Request
NFESC	Naval Facility Engineer Service Center
NRC	National Research Council
PC	Personal Computer
PHETS	Permanent High Explosives Test Site
SETA	Scientific Engineering and Technical Assistance
SS	US Secret Service
TNT	Trinitrotoluene
TWSG	Technical Support Working Group
UK	United Kingdom
USACE	United States Army Corps of Engineers
USAF/WL	United States Air Force Wright Laboratories
WES	United States Army Engineer Waterways Experiment Station
WSMR	White Sands Missile Range

BLAST MITIGATION FOR STRUCTURES PROGRAM MASTER PLAN

1. Introduction

The Technical Support Working Group (TSWG) administers, coordinates, and executes a large portion of the US Government Anti/Counter-terrorism Development Program. As a result of recent terrorist attacks against American citizens both home and abroad (World Trade Center, Murrah Federal Building, and Khohbar Towers), the Blast Mitigation for Structures Task (Task T-148F) was added during FY98. The attacks on the U.S. Embassy's in Nairobi and Dar El Salaam further demonstrate the need for blast mitigation techniques. The Blast Mitigation Program is managed by the Defense Threat Reduction Agency (DTRA) and is under the oversight of the Physical Security Subgroup of the TSWG.

2. Background

The protection of occupants of buildings from terrorist bomb attacks can be enhanced by an appropriate balance between better security procedures including the enforcement of increased standoff distances and the use of blast hardening and mitigation techniques. This program addresses only the blast hardening and blast mitigation aspects of the problem including design of blast walls, blast loading modifications, and structural hardening. Although blast mitigation methods show great promise, many of them have inadequate validation testing and some may be too costly to implement on a large scale.

The work conducted in this program will address the major issues related to preventing injuries and deaths from a terrorist bomb attack by developing cost-effective blast mitigation techniques and assessment procedures. The program pulls together both national and international agencies to generate a coordinated effort and leverage resources.

The program is broken into three major areas. The Technology and Construction Methods Evaluations area examines predictive capabilities and vulnerabilities of current construction methods in addition to screening potential mitigation techniques. The Technology Development and Application area determines the blast effects on structures and components, and develops and tests new design and retrofit concepts. Final products include vulnerability assessment methods for calculating blast effects, structural response and human injury. Design guidance will be developed to reduce these vulnerabilities. And finally, the Technology Transfer and Administration area insures that the results of this program are validated, coordinated, and disseminated to the ultimate users, and that the overall program runs smoothly and effectively.

3. Program Objectives

The purpose of the program is to develop methods to mitigate the effects of blast on structures in order to prevent injuries and death. Although structural hardening and blast mitigation techniques will accomplish this, in part, the focus is on protecting people, not the structure or other assets. Injuries and death from bomb attacks on buildings can usually be attributed to three causes: structural collapse, flying debris, and fire/smoke propagation. Since the first two represent the predominant cause of death or injury, particularly from external explosions, they will be the primary focus of the first years of the program.

A primary product of this effort will be methodologies and computer modules for predicting blast environment, structural loading, structural response, and injuries to personnel. These modules will be developed such that they can easily be imported into various vulnerability assessment, design, and analysis tools. These methodologies will be fully coordinated throughout the community so all are using consistent physics formulations. Ultimately, the final product will be the incorporation of economical design changes that enhance blast resistance and reduce the debris hazard into standard building practices.

Specific objectives are as follows:

- Establish tri-service/interagency workgroups to review and collaborate on the program.
- Develop cost-effective methods to retrofit existing structures to mitigate the effects of blast.
- Develop design guidance for new construction to mitigate the effects of blast.
- Test and evaluate commercial-of-the-shelf (COTS) products for their capability to increase the resistance of a structure to blast effects.
- Develop industry standard models for the prediction of blast effects on structures and generate computer modules for use by the government and industry.
- Define and produce joint service/agency design and assessment tools (to be defined by the user community).
- Develop simplified injury prediction models so that the benefits of blast mitigation design and retrofits can be evaluated in terms of human injury avoidance.

IV. Approach

In the first year, a review of existing methodologies and data will be initiated. Results of this effort will be used to define the state of the technology and will help focus research on critical areas. Potential mitigation materials and technologies will be identified and evaluated for incorporation into a testing program. The testing program will examine the blast response of structural components and entire structures to determine the potential vulnerabilities and hazards to personnel. Mitigation techniques will be developed and tested for both pre-construction designs as well as post-construction retrofits to existing facilities. All major issues will be worked in parallel to ensure that solutions to key problems are developed quickly.

Because of the complexity of predicting the effects of blast on structures, and the large variations in commercial construction, detailed final guidance will likely take a number of years. To ensure that results and information are made available as quickly as possible, interim guidance will be provided as the program progresses. A newsletter and technical bulletins will be published and distributed to all users to inform them of the progress of the program and to provide interim guidelines.

The blast effects efforts will leverage existing work and available information throughout the blast, shock, and structures communities. Existing facilities and infrastructure will be used to the fullest and duplication of effort will be minimized through cooperative testing. Explosive environment, structural loading, structural response, blast mitigation studies, and forensic investigation efforts will be combined as much as possible in coordinated tests supported by the Federal Bureau of Investigation (FBI), US Army Corps of Engineers (USACE), DSWA, Bureau of Alcohol, Tobacco, and Firearms (ATF), as well as international participants. To ensure the effort is coordinated and meets all agency requirements, ad hoc working groups will be established for each technology area to help guide the program. Overall program review will be provided by a committee established by the National Research Council (NRC) which will also facilitate technology transfer.

In addition, calculations, building structure component tests, individual debris hazard tests, and source characterization tests will be done. A major focus of this research effort will be testing on a full-scale representative test structure (CTS1) built without blast hardening. Retrofit methods to increase blast resistance will be applied to the structure and tested. This and other full-scale structures built during the program will be used as multi-agency test facilities to leverage resources and test requirements. The next year will see continued work on basic blast-structure interactions and the construction and testing of an structure (CTS2) similar to CTS1 but designed, within strict financial limitations, to be blast resistant. The thrust of the third and fourth years of the program will be to build on what was learned in the first two years and insure that the techniques and methodologies get transferred to the design community. Other full-scale structures will be constructed to examine earthquake resistant designs, other construction types, and to test additional blast mitigation techniques.

V. Program Summary. This program will basically have three main parts (A more detailed explanation of each part can be found in Appendix A):

1. *Technology and Construction Methods Evaluations.* This will include evaluating the effectiveness and appropriateness of current methods for predicting bomb blast effects on structures and the supporting data. In addition, standard design and construction practices will be examined to identify vulnerabilities to blast effects

resulting from applying these standards. Finally, potential techniques touted by vendors to mitigate the effects of bomb blasts will be screened to evaluate the potential for further development and/or testing.

2. ***Technology Development and Application.*** This will include a detailed high explosive test program which, in addition to basic research in blast-structure interaction, will attempt to resolve any deficiencies noted in part 1. These tests will be planned on both the responses of individual structural components and complete structures to the effects of bomb blasts. Debris hazards caused by the bomb blast and specialized circumstances such as mailroom blasts will also be evaluated in the test program. Studies of the blast loading will also be included in the testing program. This will include looking at the bomb itself (shape, composition, and configuration), its dynamics upon explosion, and the effects of intervening structures (buildings or blast walls) on the blast loading. It will then take the validated results of the research program and either develop new computer-based methodologies or modify existing design programs to take into account the potential effects of terrorist bombs on buildings. This is the most important but potentially the most difficult portion of the entire program. Blast mitigation solutions that are developed must be straightforward, economical, and simple to apply or they will not be used. Results of this phase of the program will also include development/validation of specific retrofit and protective measures designed to mitigate the effects of bomb blasts on existing structures.

3. ***Technology Transfer, Computer Support and Administration.*** Initially this program will primarily include pre-shot calculations and post-shot analysis for tests in part 2 to validate and modify predictive and design methodologies as necessary. In order to achieve the goal of coordinated data and methodologies, key organizations who would use the tools and data developed by this program will need to participate in the data gathering, development and review process. The goal will be to have this involvement conducted under the auspices of a NRC panel and to incorporate these design changes into existing building codes and standards. Some of these organizations may require funding support to participate. This task will also provide that support as well as technical support for the task manager.

VI. Program Management

The overall organization is depicted in Figure 1. The Physical Security Subgroup and the TSWG have responsibility for oversight of the program. DTRA provides program development and overall management of the effort as Task Manager for the individual tasks. The Task Manager must prepare individual task plans for each of the tasks contained herein for approval by the Physical Security Subgroup and the TSWG.

Once these task plans are approved, appropriate funds will be provided by the TSWG directly through the use of IACRO's, MIPR's, and contracts (upon availability of funds).

All task plans, briefings, and monthly reports will be generated according to the TSWG Task Manager's Handbook. The Task Manager will also ensure that the program is coordinated with DoD and non-DoD users. This will be done through ad-hoc working groups established for various technology areas.

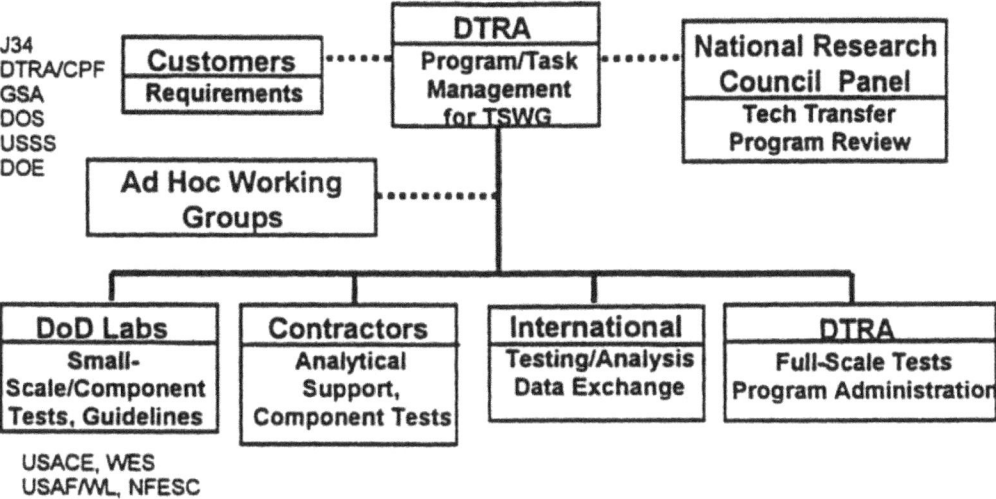

FIGURE 1. Program Management

There are many US government, commercial, and foreign government performers. The US Army Waterways Experiment Station (WES) is the lead military service laboratory under Project Reliance for Survivability and Protective Construction. They will be used for both explosive testing and methodology development.

DTRA has a considerable testing infrastructure as part of the DTRA Permanent High Explosives Test Site (PHETS) at the White Sands Missile Range (WSMR) where full-scale testing will be performed. This large facility has ample room for a number of full-scale structures and a variety of experiments with virtually no limitation on explosive yield. An opportunity exists to use the test structures to meet other agency requirements, leveraging off of the initial test objectives.

Commercial contractors will be used for much of the analysis work and for some of the component testing. Contractors will be chosen through the Broad Area Announcement (BAA) system and through standard open competition procedures.

Finally, the program will be leveraged through joint efforts with foreign governments. Programs are already underway with Israel and the United Kingdom and an effort will be initiated with Canada.

APPENDIX A - TECHNICAL APPROACH

1. Technology and Construction Methods Evaluations

A. Evaluate Vulnerability Assessment Methods/Data. Methods for predicting bomb blast effects on structures have been developed for many types of common building classes and components. For conventionally constructed buildings, these methods are often based on limited data and in some cases on no test data. The purpose of this task is to perform a detailed assessment of the analytical methods and supporting test data for blast predictions and structural response to determine which methodologies are sound and which need more validation testing and/or development. The methodologies will be researched to determine their origin and the test data used to validate the particular method. Other test data will be sought out to further evaluate the method. Where useful, first principle calculations may be used in this process. Results from this study will assist in establishing test requirements and research priorities.

B. Evaluate Standard Design and Construction Techniques. Although design standards require that structural systems be built to withstand specified loadings and other environmental conditions they leave considerable flexibility for the designer and fabricator in meeting these specifications. To prioritize the development of retrofit techniques and structural systems to be evaluated in the program, an assessment of the various typical structural types (e.g. reinforced concrete frame, steel frame, reinforced concrete flat slab, load bearing masonry), column/beam and column/slab joint designs, and field construction techniques (e.g. locations of cold joints, uniformity of concrete properties) must be evaluated. This effort will consist of a survey of design and construction practice, and analytical investigation of the susceptibility of the various different designs and construction to blast loadings.

C. Identify and Evaluate Potential Mitigation Materials/Techniques. A number of potential techniques to mitigate the effects of bomb blasts on structures have been identified (e.g., earthquake resistance retrofit methods, application of composite materials, energy absorbing materials). There are a number of vendors who claim to have products which will mitigate the effects of blast, but which are untested. These techniques and others may provide cost-effective solutions to this problem. Before expensive testing is performed, a preliminary evaluation will be made to identify the products or techniques that offer the highest potential for success. This evaluation will likely include analysis and small scale or static testing. After successful screening, these potential solutions will be incorporated into the full-scale testing program.

2. Technology Development and Application

The effort described below will leverage off of existing work throughout the blast/shock and structures communities. Existing capabilities will be utilized to the fullest and duplication of effort minimized. The program will leverage blast mitigation efforts with forensics by continuing support for the FBI in developing a better tool for explosive yield estimation and incorporating ATF vehicle debris tests into the structures mitigation effort.

A. Loading Definition. One of the primary issues that will influence the development of protection from terrorist bombs is the proper definition of the bomb environments and corresponding loads. There are several principle parameters that affect the blast loading of structures from a terrorist attack. These parameters include the location of the bomb with respect to the primary target as well as other structures and terrain, the presence of blast walls or barriers, the configuration and orientation of the charge, and the type of explosive. This effort will seek to characterize the effect of these parameters through tests and analysis. The results of this effort will help to identify advantages that may be gained by control of these parameters. They will also help to improve analytical tools that are used to determine the critical loads for design, analysis and retrofitting of facilities.

1) Blast wave propagation in complex geometries. The blast loading on structures in an urban or other area where other buildings will affect the blast pressures is not easily calculated with simplified methods. Three-dimensional hydrocode calculations are required to estimate the effect of other structures. This type of analysis was performed for Eskan Village in Saudi Arabia where average pressure reduction beyond 400 ft were found to be half of those calculated using CONWEP[1]. This showed that traditional methods might yield very conservative results for high-density areas. This task will address this issue by providing a suite of 3-D and 2-D blast propagation calculations simulating high density building layouts to determine the general trend of pressure and impulse reduction from other buildings.

2) Blast wall effectiveness. Blast walls and barriers modify the blast propagation from a bomb detonation. The Army and Air Force Security Engineering Manual (TM 5-853, AFM 88-56) contains a methodology for calculating the reduction in blast pressure and impulse behind a blast wall. The formulations are, for the most part, based upon small-scale tests conducted at WES. These formulations were never validated with full-scale experiments and are limited to certain geometries and charge sizes. For instance, the minimum height of a blast wall for a 5,000-pound bomb using these formulas will be 14 feet. This task proposes to extend the limits on the geometry of the problem through a calculational effort followed by some validation tests.

The overall effectiveness of barriers is also a function of their potential for creating hazardous debris. A recent DSWA sponsored test indicated that a reinforced concrete wall failed in a manner as to cause a significant debris hazard out to several hundred feet.

---

[1] PC-based computer code that provides conventional weapons calculations based on the equations and curves of TM 5-855-1 "Fundamentals of Protective Design for Conventional Weapons"

Tests need to be performed with various wall designs and threat levels to determine the most cost-effect ways to eliminate the debris hazard, yet still deflect the blast pressures.

B. Structural Collapse. Progressive collapse of structures has been the leading cause of death from bombings. This may continue to be a problem, particularly in urban areas where little or no standoff for buildings may be available. Column failure was the primary cause of structural collapse in the bombing of the Murrah Federal Building. A FEMA/ASCE report[2] on the Oklahoma City bombing suggests that 85% of the floor space would have been saved if the columns did not fail. This report recommends incorporating techniques used for earthquake protection (e.g. column jacketing, use of special moment frame and dual systems) into the structure for blast protection. Although seismic loading is different than blast loading, some of the methods, modified to resist blast conditions, may provide adequate protection for a modest investment. Testing is needed to verify this belief and to develop design guidance.

A number of types of structures are susceptible to collapse. Retrofit methods for flat-slab structures, load-bearing walls, and lightweight steel roof systems will be developed first. Other structures will include blast-resistant designs, moment-resistant frames, and steel frames. Specific discussions on each are contained below

1) Flat-Slab Structures

Flat-slab structures are very common in low seismic zones. Story heights can be smaller since there are no beams to interfere with overhead utilities. These structures will typically have less capacity to resist blast loads because of the weak column/slab joint. For this reason, this structural type was chosen to be examined first. If retrofit methods work on this structural class, they should work on others.

The general approach taken to develop retrofit methods is to first design and construct a full-scale flat slab test structure (CTS-1). The basic structure was designed to ACI code (This was constructed during FY98). The structure will be tested during FY99 to a blast load which will cause severe damage, and possibly collapse a portion of the structure. The remaining structure will then be retrofitted and re-tested to the same blast environment to demonstrate the effectiveness of the retrofit techniques. This effort is supported by first-principal calculations, scale model testing, and laboratory tests.

Figure 1: Typical Flat-Slab Structure

a. Scale-Model Testing

---

[2]*Oklahoma City Bombing: Improving Building Performance through Multi-Hazard Mitigation,* FEMA 277/August 1996.

Scale-model testing was conducted during FY98 to help understand potential failure modes of the structure, and to determine the structural loading. This effort is complemented with first prinicple numeric simulations of the structural response. These efforts will help insure the success of the full-scale structure test. Potential failure modes include column failure, slab failure due to uplift, or punching shear of the floor at the column. The structural loading of the column and the floor slab is complicated by the failure of the exterior wall. If the wall fails quickly, relative to the blast load duration, the blast will clear quickly and the column will have less impulse applied. If the wall fails slow, relative to the blast duration, the column will experience more impulse because the blast can not clear quickly. Another complicating factor for floor loading, is the effect of the exterior wall failure on the blast. Typical calculations assume a non-responding exterior wall.

Figure 2 shows the section of the building to be modeled, and a picture of the test article and the reaction structure. The test article simulates the center column of the building (2 stories), the floor slab (for 1 bay), and the perimeter beam. The weight of the floors of the multi-story building is simulated with a weight placed at the top of the stub column.

Five tests were conducted. The first simulated a 1000 lb bomb at 20 feet from the building. In this test the structure suffered minor damage. In the second test, the bomb was moved in to a simulated stand-off of 14 feet. In this case, some permanent deformation took place (see figure 3). There was no exterior wall in either of these tests (bare column only).

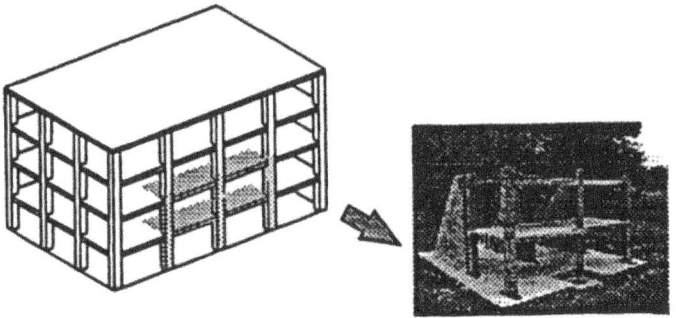

Figure 2: Scale-Model Component Test Atricle in Reaction Structure

Figure 3: Test 2, Bare Column, Moderate Damage

The third test was identical to the second except the in-fill masonry walls were added. These walls do not strengthen the structure, but delay the clearing of the blast wave so the column sees more impulse. The time it takes for the walls to break apart will affect the clearing time. Post-test examination of the structure and the pressure measurements show that the wall does breakup relatively slowly and the impulse on the column was greater, resulting in severe damage to the column. No slab uplift occurred since the blast did not enter the building until the pressures were substantially reduced (see figure 4).

Figure 4: In-Fill Walls, Severe Damage to Column

The fourth test modeled the real building exterior walls with an opening for the windows. The response of the column was expected to be somewhere between the bare column test and the test with the solid in-fill walls. The results showed a that the column received almost as much damage as the column in the solid wall test (see figure 5). In addition, the floor received severe damage. It is believed that the underside of the second floor was in direct sight of the charge and the blast was delayed to the top of the second floor by the walls. This resulted in greater differential pressures.

Figure 5: Exterior Wall With Window Opening, Severe Damage to Column and Floor

The last test was designed to maximize the uplift by leaving the bottom bays without walls and constructing solid walls on the second floor. The test results showed that the floor did receive severe uplift damage, and the column on the second floor received severe damage (similar to other solid wall test).

Figure 6: Solid Wall on Second Floor, Severe Floor and Column Damage

The results of these tests, combined with the computational effort, provide greater confidence for the full-scale tests. After completion of full-scale tests, an evaluation will be made as to the usefulness of performing more sub-scale tests to understand structural behavior in a wider variety of structural details and blast loads.

b. Full-Scale Structure Tests (CTS-1)

During FY98, a full-scale test building was constructed (CTS-1) at the White Sands Missile Test Range (WSMR) to examine structural collapse and to test retrofit methods to prevent it. Analytical studies were conducted which indicate that seismic retrofit techniques, adapted for blast loads, show promise. The types of retrofit methods being considered are composite (fiber reinforced plastic) column and floor wraps and steel column jacketing. Potential retrofit designs were developed during FY98 (Karagozian & Case contract).

A calibration test was conducted in August 1998 to provide confidence in ability of computational methods to predict structural response. Three tests are planned for FY99. The first test, which was conducted during April, was designed to cause severe damage to a column which could result in partial collapse of the building. Pre-test predictions suggested that collapse would not occur. On the other hand, if the building did not have an edge beam, or if the edge beam was somewhat weaker, collapse would occur. The test results matched the predictions extremely well. The prediction indicated a 7.5 inch deflection of the column, the actual deflection was 10.5 inches. The results provide confidence that the analytical codes can provide reasonable predictions of expected damage.

Figure 7: Pre and Post-Test Pictures of Calibration Test

Retrofits will be applied to the two adjacent columns and they will be tested to the same environment as the severely damaged column. Hopefully, this will demonstrate the ability of the methods to protect the columns. In order to develop design guidance for a variety of structural details, further testing is required and the designs need to be refined. The next section describes the effort which will result in final design guidance.

Two additional tests are planned to demonstrate structural collapse and the ability to prevent it using retrofit techniques. The first test will be a larger bomb placed between two columns. With two columns severely damaged, partial collapse will likely occur. The next test will be conducted on the other side of the building to demonstrate the adequacy of the retrofit methods.

c. Retrofit Techniques Using Composite Materials

Concurrently, an effort will be conducted to refine the use of composites for retrofitting columns, beams, walls, floors, and structural connections. The final product will be guidelines to allow engineers to design appropriate retrofits for a particular building and a given threat. It is expected that this task will be completed during FY00. (The contract is under negotiation and is expected to commence in July 99)

2) Load-Bearing Wall Structures
a. Damage Prediction Methodology

Masonry is the one of the most common building materials for both in-fill and load-bearing walls. These walls are also very susceptible to collapse because of their relatively weak capability to resist horizontal loads, and because of the large surface area to attract blast loads. Analysis methods for masonry walls have been developed for some masonry wall configurations, but only limited validation testing has occurred. Figure 8 shows an example where the FACEDAP methodology over predicts the amount of damage by a factor of 5. The graph shows a prediction of collapse while the test wall was only partially

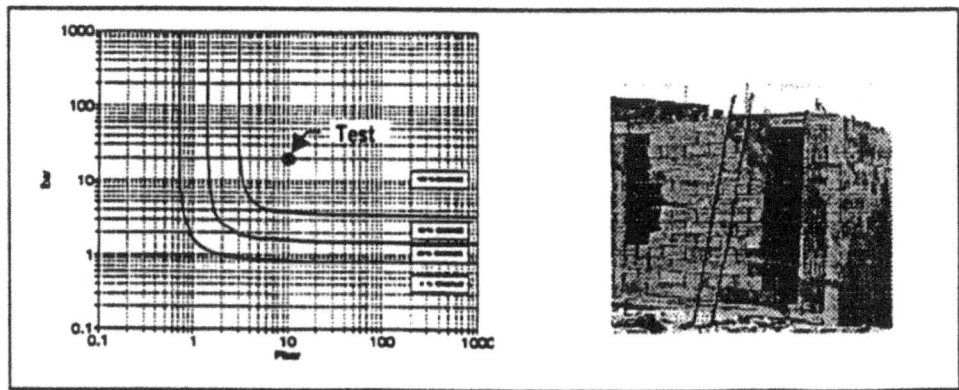

Figure 8: Masonry Wall Damage Validation Test

damaged. A design using this method would result an excessively conservative and costly retrofit. Another example is documented in the referenced report[3]. The investigators in this report discovered that the FACEDAP methodology does not predict damage well and modified the method to better model the behavior observed in the tests. Many common wall configurations (e.g., cavity walls) have not been tested and methodologies are required.

Different types of walls will be tested including reinforced concrete, unreinforced and reinforced concrete masonry units (CMU), brick, and cavity walls. Some retrofit methods to strengthen walls have been identified as having potential benefits but testing is needed to verify this. The solutions identified to-date are limited and expensive. New retrofit techniques that will be examined and tested include using composite materials, shotcrete and reinforcing, and other methods. The data from these tests will be used to develop and validate computational codes and provide recommendations of the best and most cost-effective solutions.

b. Load-Bearing Wall Retrofit Development

A multi-use reaction structure was constructed at Tyndall AFB during FY98. This reaction structure is capable of holding both roof systems and load-bearing walls for testing. Load-bearing control walls will be tested during FY99. Retrofit methods will be developed simultaneously and tested during FY99. Design guidelines will be developed by th US Army Corps of Engineers, Protective Design Center and published during the first half of FY00.

In addition to the above effort, a retrofit method was submitted during the FY98 BAA process. The proposal describes the development of a lightweight precast panel which would be attached to the exterior wall of a building. The attachments consist of shock-absorbing dampers. This concept shows promise and will be developed during FY99-FY00.

3) Roof Systems

Some common types of roof systems, such as lightweight steel bar joists, often fail during a terrorist bombing while the walls survive. There where three examples of this in the Oklahoma City bombing.

Figure 9: Roof failure in Oklahoma City Bombing

During FY98, a reaction structure was built at Tyndall AFB to hold roof test articles. The reaction structure has one removeable wall

---

[3] Validation of Component Vulnerability Curves for Unfilled Masonry Walls and Steel Joists, Karagozian & Case, TR-96-30.2, 7 January 1997.

which will provide a place to test load-bearing walls also. Two control roofs were tested during FY98 to moderate and severe damage levels. During FY99, another control test will be conducted to which is expected to achieve collapse of the roof. Retrofit methods will then be tested to prevent roof collapse. Design guidance will be developed during FY99 and will be published early in FY00. In the outyears, other roof types will be investigated.

Figure 10: Reaction Structure With Test Article - Pre and Post-Test

4) Seismic Designed Structures

Another class of structures to be investigated are structures designed to resist seismic loads. These structures are common throughout California, and are becoming more common in other areas of the world. Although blast loading is significantly different than seismic loads, seismic designs are more robust, provide more ductility, and are expected to be more resistant to blast. It should be noted that seismic design is not the same as a blast design. If one were to design specifically for blast, the resulting design would not be the same as a structure designed for seismic loads.

5) Blast-Resistant Designed Structures

It is considerably cheaper and easier to design a new structure for blast resistance than to retrofit an existing structure. Structural details, such as continuity of steel in joints, or steel in the top of a slab to resist uplift, are often the difference between progressive collapse of the structure, and limiting damage to a local area. A blast-resistant structure was designed to a threat level of 1,000 lb TNT at 20ft. The structure was designed using the TM5-1300 methodology. The structure will be constructed during FY99 and FY00 and tested during FY00. This testing will validate this methodology.

The General Services Administration (GSA) has initiated an effort to develop design guidelines to prevent progressive collapse from a bombing. The guidelines will be written so engineers without blast design experience will be able to develop adequate designs. Although blast loads are directly considered in the design process, the robustness of the resulting building will prevent progressive collapse. This effort guideline development effort will be finished during FY99. A building will be designed and constructed using the developed guidelines, and will be tested during FY01.

6) Other Structural Types

Other structural types, such as steel frames, will be examined in the out-years. Some of these types will be identified through Task 1.A, and through vulnerability assessment reports which reveal an inadequacy in analysis capability, and a deficiency in retrofit options.

C. Debris Hazards

Flying debris (e.g., glass, walls, overhead lights and utilities, other building components and furnishings) is the leading cause of injuries in terrorist bombings. Hazardous debris is generated at large distances from the detonation location and can injure hundreds of people, or even kill. Although glass is a major contributor to the debris problem, architects, building owners, and tenants like the openness of large windows. In addition, airblast entering a building can cause overhead utilities to fall, interior walls to fail, and office furnishings can become hazardous missiles.

There are three objectives in the debris problem. The first is to develop an injury prediction methodology. If preventing injuries and death is the main objective of the Blast Mitigation Program, the ability to quantify the vulnerability of facilities in terms of injury is required. Once a vulnerability is determined, the benefits of potential mitigation techniques is required in terms of injuries or deaths prevented. This will allow a risk assessment to be performed which incorporates the cost and benefits of a particular solution.

The second objective is to test and evaluate commercial-of-the-shelf (COTS) products. There are numerous vendors that have products that may increase the protection level of a facility. Unfortunately, most of the products have had little or no realistic blast testing. In addition, the tests that have occurred have not been conducted in a way to provide direct comparisons with other products. An effort is underway to test these products to generate consistent data.

The last objective is to develop retrofit methods to reduce injuries from flying debris where no cost-effective commercial products exist. During FY98, two methods were tested, and design guidance developed, to reduce injuries from broken glass and masonry walls. These methods and others will be described below.

1) Windows
a. Hazard Prediction Methodology

When performing vulnerability assessments, the metric that needs to be used is the extent of injuries and deaths of the occupants of buildings. In order to accomplish this, a hazard prediction methodology is required. This prediction method needs to take into account all blast environments and window properties, and must be able to predict the benefits of various retrofit options.

At the beginning of FY98, there were two basic methods for calculating window damage. The first is a methodology that calculates probability of breakage for a number of glass types. This methodology falls short by not being able to predict the post-breakage behavior of the glass (e.g., how far does the glass travel, what does the debris field look like). Another methodology was developed in the United Kingdom (UK) which predicts hazard levels for glass subjected to various blast conditions. This methodology is a series of pressure-impulse (P-I) diagrams (lines of equal hazard level plotted in terms of pressure and impulse from various charge sizes and ranges). Three hazard levels are described. The first is break-safe where the glass breaks but either stays in the window frame or falls inside or outside the window. The second is low hazard where the glass travels into the room but not at high velocity. The third is high hazard where the glass travels at high velocity – hits a wall 3 meters into the room at least ½ meter above floor. Although this methodology can not be used to directly predict injuries, it is a good first step toward a complete methodology. The UK Glazing Hazard Guide was developed in paper form as a series of charts.

During FY98, this method was automated and coded as a computer module for use by various vulnerability codes (e.g., it has been incorporated into AT-Planner). It has been incorporated into the WinDAS database as a standalone glazing hazard assessment tool. The WinDAS database has all of the available data on window tests, as well as a guide on how to assess window hazards and how to design and retrofit to reduce potential hazards.

Most of the data from which the Hazard Guide was developed was at small yields (100 pounds or less). There is a substantial need to obtain data for higher yields. A test was conducted during FY98 to gather data on a variety of glass types to obtain this data at a higher yield. The explosive charge size was chosen as 5,000 pounds of C-4 to ensure that the hazard level is on the peak pressure asymptote of the P-I diagram. The test was conducted at the CTS-1 structure. Forty-two windows were tested in the building and another six in cubicles. Most of the windows responded as expected with a few exceptions. The toughened glass consistently performed better than expected. In addition, the retrofit concepts performed somewhat better than anticipated. And lastly, some methods for attaching window film, in a daylight application, to the frame show promise (the frames blew in so the tests were inconclusive). This data will be incorporated into the WinDAS database during FY99. Modifications will be made to the UK Hazard Guide which incorporates this data.

One shortfall of the UK Hazard Guide is that the P-I diagrams were developed for certain thicknesses and sizes of windows (these correspond to the size of the test specimens used in the UK). Although the dimensions of most windows are reasonably close to one of the windows in the Hazard Guide, a better solution is to non-dimensionalize the guide so any size or thickness of glass can be evaluated. An effort was conducted during FY98 to examine the feasibility of accomplishing this for annealed glass. This project was a success so the plan is to extend this effort to other glass types. This effort will not be initiated until FY00 due to funding constraints.

The automation of the UK Glazing Hazard Guide is an interim step toward developing a complete hazard and injury prediction method. Glass size, shape, velocity, and debris pattern need to be estimated for incorporation into an injury model. During FY99, a method developed for the Air Force Space Command will be investigated as a base code for predicting the glass debris field. At present, this code calculates the glass shard size, and velocity for annealed glass, but has not been validated. A complete method is expected to take about two years to develop. To gather data to validate the model, the holograph system developed by the Air Force to examine fragment from penetrating munitions will be used.

The last area that needs improvement is the design of proper frames and anchorage systems. Frames and anchorage systems have been tested to some degree, but better design guidance is required

b. Retrofit Method Development

A commonly known approach to retrofitting windows is to apply fragment retention film (FRF) (e.g., mylar®) in a daylight application (FRF is applied to window in exposed areas only, not attached in the frame). When this approach is taken, the glass fails at roughly the same pressure level, but the glass travels into the room as one whole piece held together by the film, and not as pointed shards which can cause severe cuts. The filmed glass also travels at a slower velocity than individual shards at the same pressure level would have. Although daylight applications of FRF provide some hazard reduction, injuries still can occur.

One method to obtain additional protection to a window with a daylight application of FRF is the use of a "catch bar". This method uses a bar attached across the filmed window to catch the window after it breaks. During FY98, proof-of-concept tests were conducted (which were successful). A design methodology was then developed, and validation testing was conducted in Israel. The US Army Corps of Engineers Protective Design Center developed an Engineering Technical Letter (ETL) to provide complete guidance on how to design and construct this retrofit method (the ETL is currently under review in the Corp of Engineers). This method does provide an additional increment of protection, but can be overwhelmed by a large blast load. Protection levels, up to about 10 psi can be achieved with 10 mil or thicker films.

Another method to reduce glass hazards is to attach FRF to the frame. In this case, the film acts as a membrane and the blast pressures must be high enough to exceed the strength of the film or the frame if the glass were to become a hazard. This method has the advantage over a catch bar in that the it does not allow airblast to enter the building and create more debris from internal building components and furnishings. A couple of methods are being evaluated for attaching the film to the frame. One method is to apply high-strength structural silicone caulk to the film/frame. Another method is to mechanically attach the film to the frame. This was successfully tested in the Divine Buffalo 9 test. A design methodology will be developed during FY00.

c. Commercial-of-the-Shelf Products

The window industry has been involved in the development of blast resistant glazing for a number of years. Many of the glazing units are adaptations of ballistic glazing. Other products are available which include window films, curtains and shades, laminated glass, and different glass material compositions. Testing standards have been developed in the US but do not provide realistic blast simulations and therefore do not provide an adequate basis for evaluation. The wide variety of products makes comparisons between the these products difficult. Although the glazing industry has performed some blast testing, usually with very low charge sizes (e.g., .0.25-1.0 lb), there is no consistent basis to compare products. The Blast Mitigation Action Group (BMAG) was established to evaluate commercial products. This group is chaired by WES and has members from various government agencies (i.e., DTRA, Army, Air Force, Navy, Secret Service, Capital Police). The group developed a testing criteria and protocol which was presented to the glazing industry (Protective Glazing Association) for evaluation. Industry was generally pleased that the government established criteria for them to design to. The government has agreed to test industry products in a shock tube at Wilfred Baker Associates. The shock tube was chosen to perform the tests because it provides a good, consistent simulation of blast loads, and is less expensive than open-field testing. The simulated blast threat was chosen to be 1,000 lb TNT at 150 and 275 feet (peak reflected overpressures of 10 and 4 psi respectively) (a typical window breaks at less than 1 psi). The manufacturers were given the opportunity of providing up to three products for testing. During FY99, 60 products were tested. A "Yellow Pages" Internet web page was developed, and is available on-line, which lists all known manufacturers of blast mitigation products and will provide test data to government personnel with an appropriate password. Additional tests on window products will be conducted during FY00. The testing will be expanded to include door and wall products.

2) Exterior Walls (non-load-bearing)

a. Walls Without Openings

During bombing events, exterior walls often fail and become debris hazards to the occupants of a building. Failure of walls also provides for venting of airblast into the structure which can further add to the debris problem by translating office equipment and furnishings as well as non-structural building components (e.g., overhead utilities, partition walls, etc.). To protect people inside of structures, exterior walls may either need to be strengthened, or prevented from entering the building after failure by a debris catching method.

During FY97 (under DTRA/WES joint program), proof tests were conducted to examine the use of high strength fabrics to catch wall debris from masonry walls. The tests were successful and a design methodology was developed during FY98. Further tests were conducted in Israel during FY98 to validate the design methodology. The methodology was successfully validated and was documented in an ETL which was published in early FY99. Figure 12 shows the results of the test.

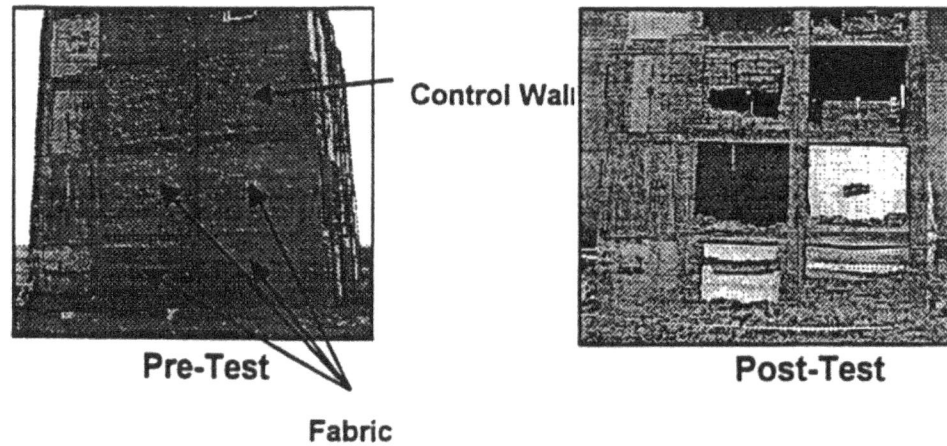

Figure 12: Fabric Wall Retrofits

b. Walls With Openings

The fabric catch method works extremely well for masonry walls without openings. Unfortunately, most walls also contain windows. Retrofitting walls with windows is a much more difficult problem. Either the window, frame, frame anchor, and wall need to be strengthened, or a catch mechanism installed which must be able catch the whole system. In most situations, the building occupants do not want the window covered.

During FY98, one method was tested using a geo-grid material in Israel and using a composite glued to the wall. The geo-grid allows some visibility through the window while the composite blanket did not. Neither method, as applied, was looked good. The geo-grid did not perform very well while the composite did.

During FY99, two contracts will be let which will provide concepts for retrofitting specific EUCOM and State Department facilities (see Technology Transfer Section). These contracts will provide retrofit new retrofit ideas which will be validated with testing.

Other combinations of efforts (composites, windows, frames) will provide other solutions over the next few years.

3) Building Internals

Interior building components such as overhead utilities, internal partitions, and office furnishings, can become debris hazards following a bombing. Such injuries were observed in the Oklahoma City bombing as well as in the Nairobi bombing where numerous deaths occurred. In earthquake prone areas, better attachment methods are required for overhead utilities. Some of these methods, and others, may provide increased protection from these hazards with minimal cost.

To develop injury prediction methodologies and methods to reduce injuries, office and housing areas will be tested in the Divine Buffalo test series starting in FY99. Testing will continue during future tests to collect additional data and examine other retrofit techniques. The product from this effort will be better debris hazard models and methods to reduce debris. Some of the methods are expected to be no-cost, smarter ways to arrange offices, to lessen the objects that may become debris hazards, and to arrange furniture to keep people away from highest airblast (e.g., keep beds away from windows). A video will be produced to describe some of these procedures.

D. Human Injury Prediction

Human injury is the yardstick by which vulnerabilities to terrorist bomb attacks are measured. When vulnerabilities are identified, numerous solutions are often available which provide different levels of protection. Since there is often inadequate funds to provide complete protection from any terrorist threat, risk assessments which examine cost and benefits must be developed. The benefits are described in terms of human injury. At present, injury models for blast focus on overpressure and human translation while most injuries result from flying debris. Models are required to estimate injuries from all major contributors.

1) Injury Case Studies of Terrorist and Other Bombings

The examination of injuries from previous bombing is useful for a number of reasons. First, the cause of injuries can be determined which can lead to improvements in buildings to prevent these types of injuries. In addition, data from bombing can be used to develop or validate injury prediction methods. One must be careful not to rely only on previous bombing events for injury prediction methods since, in terrorist bombings, the bomb type, yield, and configuration is usually not know precisely, and must be estimated based on observed damage.

To examine injuries following the bombing of the Murrah Federal Building, the Blast Mitigation Program contracted with the Oklahoma State Department of Health to complete a database of injuries sustained following the attack. Survivors of the attack were interviewed to determine the locations of the people, the types of injuries they received, and the causes of those injuries. This database was linked to AutoCAD and Arcview computer codes to provide a capability to plot the results of a query on a map of the city, and place people in proper locations is certain highly damaged buildings. The effort was completed during January 1999.

This database will be expanded with data from the Khobar Towers bombing. This effort began during May 1999 by contacting survivors with a survey. This effort will be completed during FY00. In addition, data from World War II and accidental explosions will be gathered and added to the database.

A simplified, quick running injury model will be developed using the data in the database. The method will be backed up with a more detailed method for a more refined analysis and for comparisons of various retrofit options.

2) Comprehensive Injury Model

A contract will be let during FY99 to develop a comprehensive injury model. Existing methods from the military and auto safety industry will be assembled with blast effects codes (overpressure, debris, etc.) to generate the model. After completion of the model, mitigation methods can be evaluated and compared to determine their capability at reducing injuries. This is a three year effort to be completed during FY02. The model will be put into modular format for input into various vulnerability codes.

D. Internal Detonations. A requirement has been identified to protect mailrooms and other readily accessible areas of buildings, such as lobbies, from small bombs either carried or delivered to office buildings. Techniques will be investigated to obtain different protection levels ranging from fully containing an explosion, to allowing blast venting into limited local areas. Potential methods including reinforced concrete walls, sacrificial walls, masonry walls with debris catchers, application of composite wallpaper, and others will be investigated. This task will take advantage of the full-scale test structures and will work in conjunction with the DOE mailroom initiative. The goal is to develop cost-effective methods to protect against such bombs.

1) Mailroom Protection

During FY98, a simulated mailroom was constructed in the CTS-1 structure. This mailroom used typical masonry walls found in many buildings (i.e., unreinforced 6 inch CMU, two layers of bricks, and lightly reinforced CMU). The exterior wall was designed to be light weight to blow out quickly. This type of design provides a low level of protection from a letter bomb (ref 1). The bomb was placed in an x-ray cabinet (from another TSWG project) which was supposed to contain fragments but not the gas pressure. The exterior wall was blown out and two of the interior walls fell over. The lightly reinforced wall remained in-place. The test verified the Security Engineering Manual at low level of protection. The next step is to test various retrofits and new designs for greater protection. The threat will be 2 pounds of TNT equivalent explosives. The tests will be conducted at the Chestnut Site and Kirkland Air Force Base. A reaction flame will be constructed and tests conducted during the summer of 1999. Design guidance will be developed during FY00.

Figure 13: DB-2 Mailroom Test

3. Technology Transfer and Administration

There are a number of mechanisms being used to get the technology to the users. First, each military service has a quick method to distribute engineering technology (i.e., Engineering Technical Letters, Technical Data Sheets). All products from this program will be distributed through this method. A longer-term effort is underway to develop a Military Handbook on Security Engineering. The information from this program will be incorporated into this joint service publication.

The technology developed in this program needs to be transitioned into commercial practice. A National Research Council panel was established to perform program review and develop a technology transfer strategy.